The Twentieth Century, The Age of Mediocrity

History of the Development of Nuclear Reactor Powering Systems for Freedom

by
William Wachter and John Nevshemal

John A. Nevshemal

Bloomington, IN   Milton Keynes, UK
authorHOUSE

AuthorHouse™
1663 Liberty Drive, Suite 200
Bloomington, IN 47403
www.authorhouse.com
Phone: 1-800-839-8640

AuthorHouse™ UK Ltd.
500 Avebury Boulevard
Central Milton Keynes, MK9 2BE
www.authorhouse.co.uk
Phone: 08001974150

© 2006 William Wachter and John Nevshemal. All rights reserved.

No part of this book may be reproduced, stored in a retrieval system, or transmitted by any means without the written permission of the author.

First published by AuthorHouse 4/11/2006

ISBN: 1-4259-2295-3 (sc)

Printed in the United States of America
Bloomington, Indiana

This book is printed on acid-free paper.

FORWARD

This book is the result of two nuclear engineers who developed both commercial and military nuclear opinions. They performed their efforts when the nuclear aspects of both commercial and military were on the rise. Without their involvement, the US would not be where it is, taking advantage of a major source of energy; i.e nuclear power plants that make electricity and the nuclear weapons that ended the possibility of a major international war.

This book does not foster any further nuclear war but does explain the development of a major source of energy to be used by mainly commercial producers of useable energy that does not have the environmental consequences resulting from immediate production of byproducts, i.e. environmental damage.

TABLE OF CONTENTS

CHAPTER 1
Years at Lord Manufacturing 1

CHAPTER 2
Bettis - The Early Years 7

CHAPTER 3
Initial Years of Wachter Associates 17

CHAPTER 4
Small Projects Years of Wachter Associates 27

CHAPTER 5
The Oconee Disaster & the Miraculous Recovery
Provided by Wachter Associates 39

CHAPTER 6
Fort Calhoun Spent Fuel Storage Problems Solved
So Reactor Could Start Up and Later Continue to
Operate 1973-1974 53

CHAPTER 7
The Palisades Disaster Miraculous Recovery by
Wachter 59

CHAPTER 8
Loss of Flow Test Reactor Project 1973-1976 77

CHAPTER 9
DC Cook Ice Condenser Ice Loss
Problem Solved - 1975 85

CHAPTER 10
Fast Flux Test Facility Design Review and Field
Consulting Service Including Furnishing Emergency
Equipment and Hands On Help to Get Job Done -
Saved Project, 1974-1975 89

CHAPTER 11
Spent Fuel Problem, 1976-1978 99

CHAPTER 12
Design Safety in Nuclear
and Non-Nuclear Systems 107

CHAPTER 1
Years at Lord Manufacturing

Hired as an unemployed war veteran by Dr. Zand, Director of the Vibration Lab, I was introduced to the complex world of vibrating systems. The first assignment was to read and understand Ken Brown's Paper on Vibration of Radial Aircraft Engines. It took a lot of study and finally some understanding of the applied mathematics of vibrating systems.

The more interesting part of the job finally presented itself: testing of components subjected to damaging vibrations and trying to set up in the lab the field conditions encountered by these components.

We created most of our own test equipment in the early era of vibration isolation and control. One test I created for heavy duty truck mountings had a periodic impact shock superimposed on vibration representing actual road conditions. It made such a racket that Dr. Zand made us run it only at night, leading to the first electronically controlled vibration system that monitored conditions and shut the test down if certain parameters were exceeded.

It was exciting to be working on the frontier of modern vibration testing and analysis with major discoveries in the area of dynamic damping and dynamic modulis. The immediate effect of these discoveries was the creation of new aircraft engine mountings that actually worked. The previous (WWII) engine mounts bottomed out and quit working to isolate vibrations in the first thirty (30) hours of aircraft engine operation. (No wonder the old WWII aircraft vibrated so much!)

I was assigned as staff engineer, vibration consultant to the Materials and Stresses Division of Lewis Lab. My first assignment was to make the "sand elevator" work. This was a structure that looked like a spiral staircase. Vibration applied to the spiral ramp was intended to make molding sand travel up this ramp from the first floor to the second floor, where intricate castings of jet engine parts were being made. After over a year of trying to make it work (the sand only climbed halfway up and then just jiggled), it was abandoned. I was asked to make it work. After inspecting it and watching it operate I did a few calculations and suggested they place a one inch pad of 30 durometer neoprene rubber under it. The scientists scoffed at my recommendation and I went on to work on jet engine problems. After a month my boss, F. J. Dutee, decided to try my recommendation. The sand elevator was turned on and after a few minutes it was throwing sand all over the second floor. From then on I got stuck with every little vibration problem the Lab had, such as tachometer vibration on main fan in wind tunnel, vibration of high speed camera on bearing test rigs, and critical speed analyses of various jet engines.

In the materials area I was assigned the review and development of vibration compaction of oxide materials used in advanced high temperature turbine blades. I became involved with the research lab at Wright Patterson Field (USAF). They had brought in a scientist with knowledge in the field. He claimed he had worked with the lead scientist, Dr. Heymann. Wright had set up vibration test equipment based on his (Soenhgen) recommendation. I discovered he was talking frequencies in cycles per minute and they had thought he was talking cycles per second. I had them start over, but I wasn't impressed with their German scientist and I never returned to Wright Patterson Field.

I was assigned a new project of frictionless supports for models tested in the wind tunnels. The best approach was the use of static air bearings (new idea). Using standard oil bearing analysis did not work so Dr. Pigott and I developed a new analysis. The standard analysis was based on uncompressible lubrication theory (oil being uncompressible). Using a complication of the old analysis we developed a compressible lubrication analysis. It was used in designing several air bearing supports for wind tunnel tests.

The ultra secret project of the nuclear aircraft engine became my main project. A program of evaluation of materials to be used in this program occupied my time and the work we did provided the early information in the "Corrosion Handbook" used throughout the nuclear industry.

At this time I was made the first of its kind, Nuclear Reactor Design Engineer, and was assigned the job of designing NASA's first nuclear reactor. This reactor became the Plum Brook Reactor used by NACA for material testing. During the design effort I developed early analytical techniques to handle thermal stress induced by internal heat generation in structural materials due to irradiation effects. (This was the basis for my later work at Bettis that saved the aircraft carrier reactor during the irradiation induced nuclear fuel growth panic.)

I accepted the position of Head of Applied Physics at the Brush Beryllium Company. The major project I had was the design and development of a high temperature molten metal reactor. The extreme temperatures made it necessary to develop high temperature material. Using ultrasonic vibration to improve properties and fabricability of exotic alloys, we produced improved materials for the high temperature reactor. A high vacuum continuous casting unit was developed. This unit produced the purest beryllium and uranium to date resulting in the most accurate determination of the actual density of these two metals. The density of uranium was exactly ten times the density of beryllium.

Under my direction, the world's largest hot press was constructed to make the main component of the aircraft nuclear reactor. The aircraft project was cancelled. The unit was used to make the bases of the heat shields used in the Mercury capsules. We used to call them the "frying pans", a term not appreciated by the astronauts.

The continuous casting process that we developed I tried to sell to the American steel industry to no avail. The Japanese took it and we lost a major battle in maintaining a competitive steel industry in this country.

CHAPTER 2
Bettis - The Early Years

My main project at Brush came to an end. W. W. Beaver, Director of Research, offered me a new project that I think he knew I would reject. The project involved increasing the "kill" radius of tactical nuclear weapons. At this point I accepted a position at Bettis Atomic Power Lab as a reactor design engineer.

The first step in entering Rickover's hallowed halls of Bettis was a security check and quarantine at the "Farm House". Although I had the highest DOD clearance, I still had to go through the process of review. This took almost two weeks before I was allowed through the gate.

The first project I was assigned to was the SFR (Submarine Fleet Reactor), the follow up of the Nautilus Reactor. The design was set and we were assigned the task of making it work. It still looked like a laboratory experiment similar to the first reactor.

After a brief stint in the SFR Project, Rickover's genius asserted itself. He formed a new "independent" project, <u>S5W</u>. Taking the most experienced reactor design engineers and putting them in the lead instead of the nuclear physicists, he ordered us to provide a simple rugged nuclear sub reactor capable of being mass produced. <u>We were part of the Polaris Project, the most successful project in the history of the Department of Defense.</u>

Using the Nautilus Reactor as the starting point we proceeded to wipe the slate clean and start over again. I asked for and got a drafting table and proceeded to lay out the S5W Reactor as our team, K. V. Smith, Jim Weissburg and myself brainstormed through the creative process of "original design". This was one of the most exciting and rewarding times of my engineering career.

The concept of the "module" was first conceived. (Now everyone uses this term.) The reactor was made up of an array of modules held laterally by a core barrel and vertically by unique connecting devices. I did all the initial stress analysis which was later checked by the "stress analysis section" of the project once we completed the conceptual design. My design criteria was based on possible shock loads from depth charges. It included a safety factor large enough to cover the ever upward trend of shock load requirements as actual shock measurements were made on actual depth charge tests on submarine hulls. It was interesting to note that the other submarine projects had to change their designs as the loads went up. All I had to do was reduce my design safety factors. <u>We never changed the original design.</u> Jim Weissburg used to say I actually did the S5W Reactor stress analysis in my head while waiting in line for lunch at the Bettis cafeteria. (He claimed the lady that ran the cafeteria had been jilted by an engineer - the food she served us engineers was her <u>revenge</u>.)

We completed the conceptual design in less than six weeks and now we were faced with explaining and selling the design to Admiral Rickover. Keith provided the major input to the illustrations in our design report. In addition, we had a major model building program including component models (briefcase size), component models (full size), including actual test pieces used to verify my analyses. The major model was a half scale model of the reactor in plastic. Rickover had access to all the reports and had pictures of all the models as well as a quarter scale model of the "module". Legend has it that he threw it against the radiator in his office and it didn't break. This was his usual test of any models he got his hands on.

Fortunately we got all the "creative" work done before the S5W Project was fully formed and we had the benefit of green management yet to be sent off to the Harvard Business School's "Charm School", one of the major obstacles to creative engineering through the years to the present. "A high tech engineering firm can be managed in the same way a candy company is managed."

Finally, the big day was at hand. Rickover was flying in to a meeting to start at 7 p.m. that night. All managers of the Lab and S5W Projects came with overnight kits to carry them through the agony of a "Rickover Meeting." Rickover came into the meeting room where we had placed our model of the S5W Reactor. He greeted everyone, looked at the model and said, "Best damn reactor I have ever seen. Well, what are you waiting around here for? Get to work!" At the end of this eight minute meeting he left and everyone in the room collapsed. <u>This was Rickover's finest moment.</u>

The prototype production phase fell on me and K. V. Smith went back to the research section of the Lab. (My boss told me years later that 60% of the success of the S5W Reactor Program could be credited to me and my efforts to procure and set up the manufacturing phase for mass production.) I must say that without the support and reliance placed on me by Jack Zerbe the whole program would have failed.

Lead components of the most difficult items to manufacture were presented to the industry by me. Oddly enough, the one item was the smallest component of the S5W Reactor and the other item was the largest component of the S5W Reactor.

I got the smallest item made in a local machine shop. This piece was cited as impossible to make by our lead manufacturing engineer. We replaced him with a young, wild manufacturing engineer, Warren Lester. The large component was still in the "impossible to build" stage. I visited Combustion, Foster Wheeler, and B&W. The meeting at B&W was the worst. They didn't know what they were talking about and after I asked about "weld shrinkage" in which their expert said, "sometimes it shrinks, sometimes it grows", I left the meeting in disgust. Finally I made a trip to AO Smith, one of the best pressure vessel fabricators in the world. After reviewing the component they said they could build it in quantity and gave me actual shrinkage dimensions that later proved to be "right on".

Request for quote on S5W structurals were sent to 17 companies. We received nine quotes. The low bid was from Cleveland Diesel Division of General Motors. They would not do the job unless GM was given Prime Contractor status like Westinghouse, GE, B&W, Combustion Engineering, etc. Rickover blocked this. I attended endless legal meetings to change this but we failed. Prime contractors were not liable for errors; the government covered the liability. The second lowest bid was from WF & John Barnes Company. They had AO Smith as a subvendor.

I ordered the material for the first reactor right off my layout drawing and ordered backup spare material for critical items like large ring forgings and such. A major steel strike was imminent so procuring the material had top priority.

The fabrication of the first core barrel at AO Smith required continuous monitoring and hand holding. We established inspection requirements as the work proceeded since this was a first-of-a-kind component. Early inspection criteria of subcomponents actually were based on my visual inspection; color pictures of acceptable and unacceptable surfaces were used in the inspection of the first core barrel. The workmanship was excellent. (The average experience of welders was 22 years.) Then disaster hit the project. The bottom plate assembly (base of core barrel) came out of heat treating with a 17 inch crack in the bottom surface. Could it be repaired? Who would write the repair procedure, etc.? I couldn't understand what had happened. I remembered we had a backup plate for this component, so I demanded it be tested even though its inspection papers were in order. The result of the inspection showed the grain structure of the stainless steel used in these plates was incorrect (very coarse instead of fine). With this knowledge it was possible for me to provide a repair procedure that would weld repair the crack and provide assurance that the integrity of the bottom plate assembly was not compromised. My guardian angel seemed to be watching over me. Work on the core barrel was completed on schedule.

Meanwhile at WF & John Barnes, the critical connector was being fabricated to tolerances at first believed impossible. This company specialized in tight tolerances. The man who invented the Jo-Block measuring element worked here, Mr. Johanson. I still had to mediate between Charley Cron, project manager for S5W Project at WF & John Barnes, and Ken Tess, project manager for S5W Barrel final machining at AO Smith. Each company had a different measuring system and thought theirs was the best. The other company's system was no good. WF & John Barnes relied on gauges while AO Smith relied on optics. When the fit up showdown came, the assembly of the various components fit perfectly and each side said the other guys were <u>lucky</u>.

The Twentieth Century, The Age of Mediocrity

Before we even finished the first reactor the Navy ordered nine more. To my chagrin our procurement department just multiplied the quantity of the first order by nine which really meant we had enough critical material for twenty reactors instead of ten. Before I could correct this error the quantity of reactors jumped to twenty and I was credited with unusual foresight.

This was the first and only time that a prototype of a naval reactor wasn't built and tested in the Mark 1 reactor installation at Idaho National Lab. Our prototype went right into the Skipjack submarine. The next reactor went into the George Washington, the first guided missile submarine. There were 41 of them - "41 for Freedom". I was at the boatyard when they converted the George Washington from Skipjack Class Submarine to the first Polaris nuclear guided missile submarine. We just cut the hull in half and stuck a missile section in the middle of it. I overheard two welders talking, "We just finished the final hull weld and those stupid engineers cut it apart. Those SOBs are always making mistakes."

The nuclear fuel developed by Jim Smith, the inventor of electron beam welding, was superior to the previous nuclear fuel. The process to fabricate this fuel was low cost and excellent for mass production. In addition, the fuel was more accurate dimensionally thus providing an improvement in nuclear and hydrodynamic analyses. Various tests were performed in the MTR Reactor. A final test was <u>rigged up</u>, and I do mean rigged, to provide various modifications of the fuel. The result was a stack of short fuel assemblies placed in a fuel channel and spring load upward against a rigid stop (mechanical disaster). After a year in the Mark 1 reactor in Idaho, this test was removed and found to be badly damaged. The interfaces between stacked fuel assemblies were badly cracked and the bottom was actually imbedded in the top of the fuel assembly below it. Panic took over and the metallurgists stated the failures were due to irradiation induced hydriding of the zircaloy material. A further test of this phenomena caused by one year exposure to irradiation was fabricated and ready for installation in the reactor. Naval Reactor per Rickover demanded this unit be subjected to a two week out of reactor (no irradiation) flow test. Bettis' reaction was predictable. There would be no evidence of "hydriding" failure. After two weeks the new "non irradiated" fuel assemblies showed the identical failures the irradiated fuel had. These failures were vibration impact failures, the first examples of "Impact Fretting." The stack of fuel assemblies were spring loaded upward and the hydrodynamic forces created by flow of water caused the assemblies to vibrate and impact each other. Once the "hydriding story" was erroneously announced as the cause of the "mechanical vibration failure" the damage was done and Bettis proceeded to scrap S5W Core 1, the best fuel ever made for the Navy. S5W Core 2 was made the arduous old fashioned way and the S5W Program received its only setback.

In addition to following all aspects of the S5W fabrication, I was stuck with the responsibility of following the production of "start up sources" for the S5W Project. The function of the source is to provide a level of irradiation to the reactor that will cause the instruments measuring criticality to register a small reading instead of zero and provide a continuous reading during start up. These sources were all made at Mound Laboratory in Ohio. (I hated the place, got my first real doses of irradiation at this Lab. The founder died of irradiation poisoning.) The encapsulation of the neutron source in a stainless steel capsule required insertion of source in capsule and end cap welding of capsules. The faster this operation was done the less irradiation exposure everyone got. This critical machining and welding operation was performed by two technicians. Unfortunately for Rickover these men were primarily farmers who shut this operation down during spring plowing. Rickover used to rage at the fact that the Nuclear Navy Program was at the mercy of "spring plowing".

Once the S5W Core 2 was in mass production, we came up with Core 3 that had double the life time Core 2 had. At this time it became obvious that we could come up with a core that would last the life of the submarine (never have to be refueled). However, Congress had been told that refueling, which was a big pain, could be made as easy as changing a light bulb. During this, the US agreed to furnish Britain with the S5W Reactor for their nuclear submarine fleet called the Dreadnought Class. Rolls Royce was the company handling the project, a smart bunch of engineers. They did what Bettis failed to do; extended the lifetime of the S5W Type Reactor until they had a reactor that never had to be refueled. We were told to stop extending the life of the S5W Reactors. It was making other projects look bad. At this point mass production of S5Ws supplied the Polaris sub fleet with reactors. This fleet of guided missile subs was labeled by Rickover, "41 for Freedom." It is interesting to note that we had the Russians "buffaloed" during the Cuban missile crisis. At this time there were 17 Polaris subs at sea all over the world with 272 nuclear rockets targeting 272 Russian cities and the Russians had no knowledge of where they were. With these aces in his hand Kennedy had no trouble forcing the Russians to withdraw from Cuba. The success of the S5W Program has never been equaled in the history of the Nuclear Navy. It was truly a "combat engine". It outperformed its design parameters and survived accidents like the collision with an underwater mountain. It was a great experience full of anxiety and successes the likes of which I'll never see again.

The technical accomplishments of the S5W Project laid the foundation for all future reactor designs. Certain welding and fabrication techniques are still in use today. The new generation of simplified nuclear plants cite accomplishments such as 60% as many parts in system, etc. In the design phase of the S5W Project we reduced the number of parts by a factor of <u>six</u> to one. In addition, the design was such that the reactor could be modified to fit any power requirement simply by adding or subtracting modules. The only manager to recognize that was the manager of the SFR Project. He became our best spokesman, after which he was transferred from Bettis to the Westinghouse Steam Division.

CHAPTER 3
Initial Years of Wachter Associates

A new project dedicated to silent operating propulsion systems for nuclear subs was begun. It was called the NCR Project (Natural Circulation Reactor) and eliminated the primary coolant pumps that circulated the cooling water through the reactor, a major source of noise. The cooling system operated in the boiling regime and relied on a thermal head to drive the coolant through the core and on to the steam generators. Physically, the reactor core (heat source) was at the bottom of the coolant circuit, and the steam generators at the top (heat removal point). A large analogue system was made to calculate the performance of this system. Many models were constructed to provide verification of the analogue model. A series of tests at Idaho with pumps off confirmed the NCR concept would work. Things were progressing nicely when our Lab manager Simpson decided to pull out and form Astronuclear. He took the four top technical people, Sid Krasik, Chief Nuclear Physicist; Walt Roman, Chief Design Engineer;

Walt Esselman, Thermal/Hydraulic Chief Engineer; and one other. This made Rickover so mad he took our NCR Project and gave it to G.E. Bettis never recovered from this disaster. A big picnic of the NCR Project was held in South Park. The opening of this picnic was a parade of all the managers led by Dr. Witzig, the project manager. He wore a long frock coat and held a black book up in front of him. The section managers were pall bearers of a black casket with NCR painted on the side. After the burial ceremony we had a great picnic. Dr. Witzig got up and gave a speech intended to boost our morale in which he predicted a "great future" for Bettis and urged all of us not to "abandon the ship." Four months later he quit and formed NUS with John Gray of Duquesne Lite. (So much for company loyalty!)

There was a transfer to the Research Department along with K. V. Smith and J. A. Weissburg. My first assignment was to design a small reactor to power a destroyer class ship (3500 tons or less). It had to fit in a 10 ft. x 10 ft. x 10 ft. cubicle. The reactor I invented had unique hydraulic mechanisms to operate the control rods (the element that caused the 10 ft. height to be exceeded in every design I created). Phil Ross was the general manager of Bettis Lab at this time. He asked me to bring the model of my hydraulic control rod drive mechanism to his office to demonstrate its operation. I had warned him repeatedly about the dangers of me demonstrating models of my inventions in his fancy office. He still wouldn't visit my laboratory setups so we transported the model to his office and set it up on his large mahogany table. It cycled through its operation; then a hydraulic line came off and sprayed his office with water. That was the last demonstration I ever did in his office. The preliminary design was begun and at this time four crucial limits on operation of the reactor system were set in order to keep the size to fit a ten foot cubicle. Rickover's people decided to make my conceptual design into the DIW Project and immediately all four of my limits were violated leading to uncontrolled growth in the size of the reactor. To accommodate this size, the destroyer ship grew from 3500 tons to 7000+ tons and became a frigate. I gave up in disgust and rejoined the Research Department.

A major problem was emerging that affected the nuclear fuel of all submarine and surface ship reactors. I was assigned to aircraft carrier reactor projects. The problem was irradiation induced growth of the nuclear fuel itself. Even though the growth occurred during operation that exceeded one year in a nuclear reactor, Naval Reactors (NR) issued an edict that we must redesign nuclear fuel to accommodate this growth (about one percent) elastically. Submarine Project obeyed this command like robots and doubled the cost of the manufacture of their fuel assemblies. I exploded and made such a fuss NR backed down if we could provide an analysis to show this growth was not a problem. We formed a team to do the analysis - Dr. Harbaum, computer expert in the field of hydraulic analysis; Jim McCauley, physical metallurgist; and myself. I had to educate the stress analysts in the area of plastic-creep analysis based on my NASA experience with the analysis of turbine discs. I enlisted experts from all over the world. The best were from Australia. Part of the educational process was the introduction of creep/plastic terms that no one at Bettis had ever heard of. After a few weeks our team wrote the first plastic/creep computer code (called FESTR) at Bettis. We ran this program on the debug classification at the computer center. It ran the first time and I ran the analysis that saved a twenty million dollar reactor core. This was probably the only time I know of that a computer code actually saved money. A series of computer codes evolved from FESTR, but it was dropped because of its name. Dr. Harbaum left Bettis and became Director of Research at Proctor & Gamble, computerizing all their chemical processes over the next few years. I returned to the advanced reactors section of the Research Department. The irradiation induced creep analysis effort created computer codes, created from my original analysis, formed the basis for most of the future analyses in this area. I determined that zircaloy also grew under irradiation and its interaction with the nuclear fuel complicated the overall analysis. Years later I warned of zircaloy growth would lead to growth of fuel rods in fuel assemblies. A clearance must be provided at the top of the fuel assemblies to permit the zircaloy fuel tubes to grow. Many early fuel assemblies failed to provide clearance and fuel rods bowed and caused problems.

An advanced destroyer reactor project was formed and I was assigned to it. We were eliminating the control rods completely in this design. The control concept involved the use of "neutron flux traps" first invented by Henry Kirk, the most creative nuclear physicist at Bettis since Sid Krasik went to Astronuclear. Several innovative designs were invented by our team.

Interest in the project waned and we were assigned to a new project that Rickover initiated under the table without any approval. The project later became known as the LPR Project (Large Power Reactor). My main job was to convince the Admiral that fuel rods were as safe as fuel plates. A series of "what if" tests were performed to show the fuel rods were as stable as fuel plates. In addition, rod geometry increased the amount of fuel per unit area by at least two to one over plates. The next phase of this job was to develop fuel rod support systems.

I set up a series of basic flow tests of fuel rods supported in various ways. Using my NASA experience I built a "wind tunnel" at Boehmer Heating Company. Dave Boehmer did most of the work of metal bending and construction. The wind tunnel was used to perform the basic oil droplet flow tests on fuel rod arrays to obtain basic knowledge on fluid flow through rod arrays. K. V. Smith and I invented the first of the spring-stop (spring pushing rod against rigid contact) grids that was the basic fuel rod support system used throughout the nuclear industry. This was my only real disastrous invention. It was based on an entirely wrong theory of fuel rod support and I have regretted it. A worse version of it was invented by Westinghouse Commercial Nuclear Power and this was the version that everyone in the nuclear industry copied. The flaw in the analysis of this problem was the assumption that clamping force at the rod support was the critical factor. A year of test results performed in the high temperature pressurized water test loops at Bettis provided me proof that the real critical factor was the spring constant (stiffness) of the support and that this stiffness only had to be a factor greater than the stiffness of the rod span between two supports. This fact combined with the fact that the grid <u>force</u> went from one hundred percent to less than twenty percent over the life of the grid due to irradiation stress relaxation, while the stiffness never changed (modules of elasticity never changed). Why people have been unable to do the analysis performed by Dr. Akey to prove the above facts is a blot on the capability of nuclear-mechanical engineers to understand and perform basic vibration analysis. (I noted this weakness at the Westinghouse Commercial Nuclear Power Division when I was a consultant to them in the late sixties.) The grids had an additional defect common to all modern day grids. They all marked the fuel rods at their supports. This violated one of many restrictions imposed on the fuel rod. At this time I invented the spacer collar shown. The spacer collars were placed on the fuel rods and acted as lateral support at intervals along the rods. The collars were fixed to the fuel rods and contact adjacent collars, thus eliminating any chance of fuel rod rubbing at the fixed grid supports of the previous design.

The test programs to confirm our fuel rod support system were as follows:

1. Fretting and Wear Test - Battelle Columbus

2. Materials Review - Argonne Lab

3. Analytical Modeling and Formulation of Dynamic Models of Nuclear Fuel Assemblies - University of Pennsylvania Franklin Institute

The best organization in this program was Battelle Columbus (far superior to Battelle Northwest). The job given to Dr. Zudans of the University of Pennsylvania was beyond his capabilities and the capabilities of his organization. I gave them this part of my program because I was ordered to do so by management at Bettis. A former Bettis manager was made head of Franklin Institute.

Continuous flow testing of various grids confirmed that the requirement that the grids would not mark the fuel rods could not be achieved. The only support system meeting all the requirements of the Admiral was the spacer collar system. Flow tests of this system covering the extreme limits to be encountered in the Large Power Reactor (LPR) further confirmed the ability of the spacer collar to perform for the life of the nuclear fuel.

All of the above work, including designing and building test nuclear fuel rod assemblies was done <u>without authorization</u> of the project. When the project was authorized and funds made available, Rickover's staff descended on us demanding to know where all the reports and test results were. We said, "Your boss has all the original reports. We were <u>ordered</u> not to keep any records until the work was <u>legalized</u>." We were legal from this time on (no longer risking federal prison time).

The first major project to be initiated was a full size fuel assembly test to be installed in the center of the Shippingport Reactor. Rickover wanted this assembly installed in less than nine months. Bettis Management said it would take eighteen months. Rickover called them a "bunch of old women" unable to do anything in less than a year. This got Phil Ross madder than "hell". He called me into his office and wanted to know "could I meet Rickover's schedule." I said only if I have unlimited access to all the machine shops, etc. This means stopping any work in progress and putting my work ahead of everyone else. He agreed and I proceeded to make a lot of enemies. We named the project "Special PWR Cluster SABRE Project." The original grids marked the dummy fuel rods during high velocity flow tests. We had to replace them with spacer collars. The spacer collars passed the high velocity flow tests. There were sixteen different fuel pellet loadings in the 48 rod array of the SABRE assembly. Each rod had an instrument to measure fuel pellet growth. I designed a special shipping container to isolate the SABRE assembly from shock and vibration loads during shipping to the reactor site. Excessive vibration loads on this highly instrumented fuel assembly could affect the instrumentation. Prior to shipping SABRE to the Shippingport site, I took the shipping container on a "dry run". Truckers told me the route from Bettis to Shippingport was the worst in the nation for road shock and road vibration. I had special measuring devices that I recorded enroute. (I rode on the trailer in the dead of winter - it was a cold ride.) The mountings on the shipping container eliminated shock and vibration loads to the container. We installed the SABRE Fuel Assembly in the reactor <u>seven months</u> after the SABRE Project was begun.

While I was occupied with the SABRE Project a series of small fuel rod tests were installed in Idaho Lab Mark

The Twentieth Century, The Age of Mediocrity

1 reactor. Spacer collars were abandoned even though they outperformed all the Grid tests because of the notion that "clamping force" was the most important requirement of fuel rod support systems. The whole nuclear power industry went astray here and never got back on track as far as fuel grids were concerned. My years of testing was ignored during which I proved that grid material was not a major element in grid design. We successfully tested zircaloy grids and zircaloy spacer collars to demonstrate this fact. The "grid task force" (they wasted millions) went with ASTM350, a material that showed the lowest irradiation induced stress relaxation (important only if you believed the gripping force myth). The frightening thing about ASTM350 was its brittleness and the fact that it could crack and fail if stresses got too high (no plasticity).

I returned to the actual reactor design group and we developed the "moving fuel control concept." In the LPR reactor design we were forced to include shutdown rods as a backup to "moving fuel". This led to a complex module with an even more complex control rod drive system. A quarter scale model of this compound mechanism was built and the day before it was to be shown to Rickover the ball nut drive failed and .030 inch balls fell out. An overnight search recovered all but one ball and the unit was reassembled and operated for Rickover that day. It was the intention, since Rickover and California's Governor Brown (the older) were friends and Rickover knew there was a three billion dollar "water fund" available, to build a 500MW Power Reactor (LPR) to be located at one of two sites. The first site was in the Ta'hatchapl Mountains in an earthquake zone. The second site was on the Pacific coast where we had to worry about tsunamis (giant tidal waves). I was given the task of analyzing the two sites and proposing design requirements for the power plant.

After a year of reactor testing, SABRE was removed and sent to Idaho. I was given the job of disassembling the unit. The Rickover edict was no scratching the fuel rods. Using the underwater milling machine, we cut off the top of the channel the fuel rods were in and then cut the bands holding the spacer collar supported fuel rods. Again the spacer collars performed flawlessly. The project was downsized to a Shippingport Experiment when Governor Brown lost his election.

I moved out of the project and resumed designing military reactors. Things were going down hill at Bettis. After a life of designing plant fuel-rodded submarine reactors that no one was interested in, I quit and formed my own company, Wachter Associates.

CHAPTER 4
Small Projects Years of Wachter Associates

In October 1967 I left Westinghouse Bettis Lab and formed Wachter Associates. The first job I had was consultant to John Taylor, current head of Westinghouse Commercial Nuclear. A disastrous failure of the Selni Reactor in Italy was the reason for being hired. I was given the job of redesigning fasteners that held a three piece thermal shield together. The original design failed along with several other parts of the reactor. When I arrived Paul Campbell, a senior design layout draftsman at Bettis, came up to me and said "Thank God you're here, there isn't a design engineer in this whole division. We draftsmen do all the design work." After surveying the job assigned to me I went to John Taylor and told him no one could fix the thermal shield. He was wasting Westinghouse's money. I was kept on the job and came up with several "band aid" fixes. The term came from Bernie Langer who was a senior advisory engineer for the division. Most components were underdesigned and too flimsy. The thermal shields were a prime example of that. I reviewed the component at Selni and stated this thermal shield was "too flexible and free to move", a common fault that would plague the industry for years. We ultimately removed the thermal shield and cut it up in pieces that fit spent fuel shipping containers and shipped the pieces off for disposal.

While at Westinghouse I was retained by KPA Inc. to work at the Rockwell Research Lab to form a team with Dave Pettigrew, the inventor of most of Rockwell's small tools sold around the world. We designed a drilling machine based on a new process that could drill a hole through stainless steel at the rate of 50 inches per minute. (A steam generator tube sheet is 30 inches thick and has 3000 holes drilled in it.) The drilling machine we created was the first major advance in the drilling process in over 200 years. Major concern was the vibration of the frame structure because of the high drilling loads. The testing of this machine was a gigantic success. There was no vibration and little sound. It was so quiet a signal light had to be added to alert the operator when the machine was operating. The success of the prototype machine was phenomenal. Rockwell Tool Division abandoned this system because it was four inches taller than the height limit established by the company!! No wonder Rockwell faded as a corporation like the lack-of- vision steel industry and other industrial segments of the United States.

Wachter Associates consulting services were requested by Budd Manufacturing. We were called on to review the Metroliner Project being fabricated and tested by Budd. Westinghouse was supplying the electric power to operate the metroliner trains. Westinghouse and GE were given contracts to supply on-board electricals including drive motors and controls. Building of the engines was taking place in one of Budd's assembly shops. Jim Weissburg was sent to witness what was being done at this site. He reported that the work being done on the engines was complex and needed better lighting, much cleaner area and overall upgrading of the facility. He said, "It's like trying to build a steam locomotive and a color TV in a dirty locomotive shop." The complex wiring of the motive system/controls and complex signaling and safety systems required 280,000 electrical connections with associated wiring. The wire used was railroad standard black insulated wiring (no color coding so necessary in complex wiring systems). In talking to the workmen doing the intricate wiring and connection of components Jim observed how poor the lighting was. The workmen told him that errors in the connecting work ran as high as 30%. Upgrading the lights in the work area resulted in reducing the errors to less than 1%. We demanded the black wire be replaced by color coded wire to make it possible to maintain the system and repair it in a reasonable manner. Budd hired six MIT student engineers for summer jobs and had them trace and replace the wiring over the next summer.

When the prototype metroliner made its first run on a section of the New York to Washington line, everything went wrong. All the power transformers along the line blew because Westinghouse engineers designed them for "steady state" operation using a fancy computer code. This error could have been avoided if they had called in Bernie Langer. He was originally involved in streetcar power lines as a very young engineer. He would have told them you don't design transformers on a train power line for steady state as if the system was a line to light homes and factories. The train power transformers are designed for "transient state operation." Every time a train goes by, the transformer sees a surge in power required. This mode must be designed for. Meantime on board the train GE engineers were replacing circuit boards in the engine and cars as they failed. The circuit boards were "hard" mounted to the frame of the car and had electric contacts with twenty thousandths of an inch gaps that chattered and arced continuously.

After many months the major mistakes were corrected, but system maintenance would always be a problem. Upon completion of our consulting work our report noted that metroliner trains could be kept running by cannibalizing parts from spare trains. This was how the trains were kept running in the early years.

While the Metroliner Project was occurring I received a contract to do a design review of B&W's nuclear fuel assemblies. Major concern was the effect irradiation induced creep/stress relaxation would have on this component. The analysis confirmed the ability of the grids to take irradiation induced loads. An emphasis was placed on maximizing the elastic deflection of the grid springs (0.015" or more) to minimize the probability of an open cell (fuel rods can rattle in the grid). This is the only real dynamic failure observed in my two years of fuel rod support testing at Bettis. During this time I had complete control over all the flow test loops. B&W

asked me what constituted conditions leading to failure. The answer was "any rod having two consecutive open grid cells (looseness) will result in an impact fretting failure." The statistical probability of this happening is magnified by the spring/stop designs used on all modern fuel grids. Almost all nuclear fuel rod failures have been the result of rods impacting against rigid stops. B&W set up a flow test of a dummy fuel assembly having a fuel rod with two consecutive open grid cells. After a twenty hour flow test the rigid grid stop on the top grid of the two open cells had impact fretted a hole through the cladding and part of the fuel pellet within the cladding. This result shook people up so much that this test was never documented.

At the winter meeting of the American Nuclear Society I picked up two large contracts. The first was a contract to design a long life nuclear fuel assembly for NUMEC, a new company founded by Bettis and GE engineers to make nuclear fuel assemblies for the nuclear electric utility industry. The design created by Wachter Associates was based on thousands of hours of flow tests in and outside reactors. The main object was to minimize the possibility of impact fretting, the only failure mechanism observed in all the tests. The fuel grids had no rigid contacts and relied on grid springs only. This was the only fuel assembly designed for 60,000 MWD/ton burnup, which is twice the designs available at the time. Several of these fuel assemblies were tested in the Haddam Neck Reactor. Unfortunately, NUMEC was sold to Richland Oil and the manufacture of nuclear fuel assemblies was discontinued.

The second major contract at this time was a design review of the Combustion Engineering (CE) nuclear reactors. This review revealed major weaknesses in the thermal shield attachments to the core barrel. It was recommended that the bottom edge of the thermal shield be rigidly attached to the core barrel by means of eight blocks, thus fixing the lower end of the thermal shield that was free to move radially the way it was mounted. This modification was never implemented and a few years later the component came loose and caused great damage to all CE reactors of this type. I even predicted the time of failure of each reactor. The lateral restraint of the thermal shield was a set of jack bolts mounted in the thermal shield and set against the core barrel. At first CE specified a jacking force obtained by 50 ft-lbs of torque applied to the bolts. I pointed out how unreliable this method of loading was and got them to obtain loading by measuring deflection between thermal shield and core barrel. The deflection proved to be too little and irradiation induced stress relaxation occurred allowing a gap to occur with the major failure mode of <u>impact</u> <u>fretting</u> which caused the thermal shield to fail. Large chunks of the shield in the support areas broke off and the thermal shield came loose. I warned all utilities owning this type of reactor (Millstone 2, St. Lucie, Maine Yankee, Fort Calhoun). Northeast Utilities informed me that Millstone 2 was okay. They had just inspected it. I told them to go back and look again. They discovered the thermal shield had failed as I predicted.

Westinghouse contracted with me to provide a means of overcoming steam jet forces in a possible break in the horizontal pipe out of the top of the steam generators at Indian Point 3. These forces would overturn the steam generators if an earthquake occurred at the time of the pipe break. The answer was a steam thrust turner as shown in Figure 4-4. A pipe break jet force occurring in the horizontal run of the steam pipe would be turned downward by the left pipe elbow. My engineering judgment decided the worst pipe break would occur at 45 degrees in the left steam elbow. Westinghouse engineers forced us to analyze the pipe break at every ten degrees raising the cost of the project by a factor of ten. <u>45 degrees proved to be the worst break</u>.

When Westinghouse formed the Nuclear Barge Project I was a consultant on <u>mass</u> producing containment structures in a large boatyard type setting. The object of this project was to mass produce nuclear electric generating plants in a "factory setting" instead of "field setting," thus greatly reducing costs and at the same time greatly improving the quality (safety) of the final product. Upon completion, the barge mounted unit would be towed to a dredged out coastal site and permanently lodged in this site. This was a good idea if we did what Phil Ross recommended in his paper "Implications of the Nuclear-Electric Economy." The French took his advice and have a Nuclear Electric Economy. The barge project proceeded toward the final design stage before the project was cancelled. I had developed a containment structure that cut the total welding of the steel shell in half simply by doubling the width of the plate proposed by Newport News Shipbuilding which was part of our team. They turned down this proposal using the "We use our standard narrow plate because we have experience with it." Then they hit me with the biggest technical mis-statement I've ever heard. "We can weld two narrow plates together for the same price you can get a single wide plate." This weld had to pass a series of inspections, including X-ray. Besides you can't find anyone who can roll a wide plate. (The shipyard in Florida across the bay from our "factory" site was able to roll this wide plate.)

Wachter Associates was hired by General Dynamics to evaluate their gas cooled reactor designs. The main problem is always the steam generator. They had made maintenance of generator tubes easier and cited this approach as a big improvement. Our comment was the fact that the tubes most likely to fail were the hardest to maintain. Our overall design review of their latest reactor design was not encouraging. Some people said I had a great deal to do with General Dynamics selling this division to Gulf. Later I was recalled to help design the fuel assemblies for the gas cooled fast breeder reactor. This was a government supported program. The government demanded Gulf hire one of the "Big Four" - Westinghouse, General Electric, Combustion Engineering, Babcock & Wilcox - to help design the fuel assemblies since they were similar to the PWR fuel. Gulf refused to do this, instead they told the government they were hiring Wachter Associates to design their fuel assemblies. The government said that I was as good as any of the "Big Four" so Gulf called me "one of the big five." The design of the Gulf fuel assembly was based on my previously developed design criteria. After initial testing the gas cooled fast breeder was cancelled.

I was concerned about the only gas cooled reactor being built in this country - Fort St. Vrain. Those people had little experience with helium systems and were headed for trouble. I asked, "How are you going to heat up the gas at start up?" They answered, "By pumping it." You can't; gas is entirely different from water. Heat by pumping water is much greater than heat from pumping gas. The fans pumping gas during reactor operation won't heat up the gas during startup; <u>high-friction fan blades</u> are needed. After several attempts failed to heat the gas, they changed fan blades as I told them. Next came the problem of degassing the reactor system. My experience with high temperature helium systems in the Manhattan Project told me it would take weeks to outgas the reactor system. After several attempts to start up the reactor after a short heat up period (fan blades failed), the operators got smart and outgassed the system for <u>weeks</u>.

In this same period I had several small nuclear fuel design jobs from several companies entering the commercial nuclear fuel business. Everyone copied Westinghouse even though I did have a chance to improve on their <u>bad design</u>. CE came out with the first zircaloy grids to support nuclear fuel. Westinghouse went to all CE customers and predicted early failure of this fuel. I went around after Westinghouse telling the utilities the fuel wouldn't fail - Westinghouse didn't know what they were talking about. CE later thanked Wachter Associates for supporting them. They were told that their fuel wasn't any worse than Westinghouse's; <u>neither designs were any good</u>. The point I made with the utilities was that grid material didn't matter, zircaloy is as good as high strength 718 inconel since strength had nothing to do with it. At this time <u>major</u> <u>projects</u> were being awarded Wachter Associates and another phase of my history was beginning.

I got a call from Ellis Cox, Bettis' project manager for the Aircraft Carrier Reactor Project. He had just become president of B&W, (B&W was on the rocks, losing 21 million dollars a month). I met with him and his top vice presidents, (fortunately George Kessler, VP of Engineering was one of them, very important later). He informed his staff that I was the type of consultant he wanted. He further stated that he did not like me; I was mean and hard to get along with, <u>but</u> I was the best engineer he had ever worked with and my decisions were always right on! He told me there were four problems he needed help on. The first involved nuclear fuel; the second steam generators; the third coolant pumps; and the last, coal pulverizers at the Cardinal Plant. He asked me which problem did I want to work on. I told him since I was supposed to be the world's expert on nuclear fuel I would take that problem.

Upon returning home to Pittsburgh, the phone was ringing. Ellis was on the line. He said I was going to work on the coal pulverizer. I said I don't even know what the thing looked like.

He said that it didn't matter, it was a vibration problem and I was an expert in vibration, (he remembered my work at Bettis). By this time Jim Weissburg was working for me. We made the trip to the Cardinal Plant on the Ohio River in West Virginia. There were eight of these monsters, each grinding 100 tons of coal an hour to a fine powder. The vibration actually caused the ground to shake in the area of these machines. I called Ellis Cox and told him he could perform seismic vibration tests on the ground at Cardinal. He cursed me and said, "Fix it!" We proceeded to reduce the vibration by redesign of the pulverizers. This project made me realize that in the long run, nuclear power was the final answer to energy in the 21st century. These pulverizers consumed a mountain of coal every day which was turned into a mountain of fly ash. One 1000 MW coal plant will produce fly ash that would fill a football stadium to a depth of eight feet in less than a year. Handling this waste from 500 coal plants is a task beyond comprehension. The alternate energy (solar and wind) would occupy significant areas of the earth and cause environmental problems not even considered (wind farms wipe out migrating birds).

Upon completing the Cardinal Project for B&W, I received notice from the U.S. government that I was a delegate to the first conference on nuclear power at the United Nations. I commuted daily from the Poconos to the UN building in New York during the conference. With my fresh zeal for nuclear power I talked to a lot of delegates from all over the world. To support my discussions I relied on Phil Ross's "Nuclear Electric Economy" paper. The only people who got the message were the French. U.S.'s regulatory bureaucrats and anti-nukes did us in. At first I drove my old station wagon in and parked right in front of the entrance to the UN building. The second day I came out and my car was surrounded by large black limousines and half the New York police force. President Mizutu of Zimbabwe was there. I was stuck in my parking spot for over three hours. From then on I parked at the far end of the building. The last day, the UN gave a dinner at the UN dining room.

CHAPTER 5
The Oconee Disaster & the Miraculous Recovery Provided by Wachter Associates

Early in March 1972 I received a frantic call from George Kessler, Vice President of Engineering, B&W Corporation. A major Oconee reactor test (hot functional) had failed disastrously and no one knew what to do. The Nuclear Division at Lynchburg was in a state of shock, employee morale was measuring zero, and engineers were getting crank phone calls, as well as nasty letters.

B&W Headquarters in Barberton, Ohio, chartered a plane and flew me to the reactor site. I performed an inspection of reactor internals. Everything tore itself apart and loose pieces ended up on the top of the B steam generator where they caused 3300 tube sheet welds to fail. I demanded an account of what happened. After about six hours of hot flow test (thank goodness the nuclear fuel was left out for this test), a technician heard a noise. He called his supervisor; after a while the supervisor called his boss who came and listened. He called his boss and finally after hours of noise listening, someone shut the test down. I raised hell with everyone involved in this comic opera and told them, "If you hear <u>anything</u>, <u>shut it down</u>"!

The Oconee reactor design was the result of a flow test interpreted and implemented by impractical fluid flow engineers. Instead of allowing the coolant flow (water at 500°F and 2000 PSI) to mix and flow down the outside of the core barrel, they essentially put the equivalent of shower head nozzle baffles on each of the four inlets creating four jet streams flowing down the core barrel/thermal shield annulus. The four jet streams intersected under the bottom of the core barrel causing severe whirlpools in this region. To overcome this problem huge mixing vanes (similar to vanes on a meat grinder) were installed on the reactor vessel bottom. These vanes were five feet tall in the bottom of the reactor vessel tapering to one foot up the side of the vessel. Although the reactor internals would have still failed if these flow directors were not in place, the failures would have been less severe.

I immediately implemented a program of inspection and measurements to determine the severity of the hydraulic forces acting on the reactor internals. The major failure was the thermal shield supports, thus allowing the thermal shield to rock and impact the bearing pads at the top of the thermal shield. This resulted in a classic case of impact fretting and part numbers were transferred from one component to another in the area of impact fretting - further proof the failure was impact fretting. This 50,000 pound cylinder (thermal shield) of stainless steel was attached to the core barrel by means of eight 1" radial dowels held in place by 1/4" fillet welds. These dowels failed and wore conical holes in the thermal shield and core barrel, allowing the thermal shield to rock back and forth on the core barrel and wear .060" off the bearing pads at the top of the shield that restrained it against the core barrel. I called George Kessler (B&W Barberton) at this time and told him I could fix the Oconee Reactor so it would work.

Previous experience with thermal shield dynamic motion and subsequent fastener failures confirmed that the Oconee thermal shield failed by first breaking the lock welds on the radial dowels at the bottom of the shield combined with non-preloaded pads at the top produced an unstable dynamic motion of the thermal shield. (Rocking and rolling about a fixed point on the bottom edge of the thermal shield.)

Examination of the wear at all points of failure confirmed this conclusion. A typical impact fretting pattern appeared at the top restraint points of the thermal shield. Fortunately, the length of time the shield was subjected to full flow conditions was relatively short to prevent the ultimate failure that would have resulted in major damage to the core barrel assembly.

During my return by charter flight to Pittsburgh I conceived the preferred fix. It consisted of turning the shield upside down, drilling 150 1" holes and tapping them at 8 threads per inch in the rim after machining it flat, machining the lower barrel to give a slight interference fit with the shield, and then bolting the two pieces together. If necessary, the upper lip of the thermal shield should be clamped to the barrel by means of a <u>flexing</u> clamp. Calculations at this time showed that the upper clamp was <u>not needed</u>. B&W put a brake shoe support at the top, leading to bolt failure. Upon discovery of the wear patterns at the dowel locations and the bearing pad locations, I proposed a test that would provide a conservative upper bound determination of the forces acting on the thermal shield. This test consisted of inserting the dowels in their respective holes in the core barrel flange and rocking them in all directions while measuring the motion with a dial indicator gage. By simple geometry, it was possible to determine the maximum rigid body motion of the thermal shield. Subtracting the final clearance existing at the upper pads provided the elastic deflection of the shield at each pad location. This elastic deflection was used in an existing structural flexibility computer code to determine the upper bound forces acting on the thermal shield. A previous calculation based on previous fluctuating pressure loads experience in other thermal shields resulted in an overturning force of 200,000 pounds. The calculation based on the dowel rocking test gave a force of 114,000 pounds.

B&W Lynchburg continued to go its own way. Their design called for <u>16</u> 1" bolts with 8 threads per inch instead of <u>150</u>. In support of their design they called in a collection of consultants from all over the world to review the problem and recommend fixes. After an all day meeting Dr. Den Hartog, a top vibration expert from MIT, got up and recommended B&W instrument the reactor internals and put them back together, then rerun the hot functional flow test until failure and determine the cause of failure from the test data. After

The Twentieth Century, The Age of Mediocrity

Duke officials recovered from heart attacks I got up and stated "If you make it strong enough, it will work. I can do that!"

After the meeting I was saying goodbye to my old friend Dr. Den Hartog when Harry Mandel who formed MPR Associates after resigning as Rickover's right hand man, came up to Den Hartog and said, "I don't know if you remember me, but when you were a consultant to Admiral Rickover on the Submarine program . . ." "I never worked for that monster in my life," said Den Hartog. "I worked for my old friend Bernie Langer at Bettis Westinghouse." This episode cleared the way for me to use Bernie Langer as my chief consultant on the Oconee Project. I used him to intimidate the Lynchburg engineers when they interfered with my work.

Bernie Langer developed a series of analytical tools for designing components subjected to high vibratory forces. We rewrote this collection of formulas in a more general form and called this document "Guidelines for Design of Reactor Components Subjected to High Velocity Coolant Flow." Using his analysis provided a means of determining whether reactor components would fail or not. Every component that failed during hot functional flow tests of the Oconee Reactor failed to pass the "Langer Criteria." All the redesigned components that passed this criteria did not fail during extensive hot functional flow tests of the Wachter designed Oconee Reactor.

The final design of the Oconee reactor internals was predicated on the violence of the failure components experienced during the first 20-hour hot functional flow test. The major components redesigned are the thermal shield, guide tube nozzles, in core guide assembly, lower grid flow distributor assembly, and the surveillance sample holders (similar to the thermocouple guide tube.

A. <u>Thermal Shield</u>

The redesign by Wachter accepted by B&W Lynchburg involved turning the shield upside down, drilling 100 1" holes and tapping each at 8 threads per inch in the rim after machining it flat, and machining the lower barrel to give a slight interference fit with the shield, then bolting the two pieces together. B&W added unnecessary brake shoe clamps at the top of the thermal shield (a 5 to 10 degree temperature difference imposed five million pound load on the bolts). Analysis and final tests showed the top support was not necessary.

B. <u>Guide Tube Nozzle</u>

Another high damage failure (27 out of 55 nozzles failed) that destroyed the tube sheet on the B steam generator. This component was redesigned to be ten times as strong as the original design and has never failed in any of the 12 nuclear reactors of this design. It satisfies the Langer Criteria.

C. <u>In Core Guide Assembly</u>

This component was also strengthened by a factor of ten and meets the Langer Criteria.

D. Lower Grid Flow Distributor Assembly

This component was reviewed by W. Wachter and B. Langer and a static analysis was performed at Langer's request. The results indicated problems could occur during heat up and cool down (thermal differentials) if the support plate was bolted to the bottom ribs. These bolts were removed.

E. Surveillance Tube

This component was evaluated using the Langer Criteria and found to be inadequate hydrodynamic critical natural frequency = 98 hertz, structural natural frequency = 66 hertz. We proposed adding an intermediate support raising the structural natural frequency to 200 hertz. The B&W fix was to thicken the walls of the surveillance tube (raising the frequency to 100+ hertz). This unit failed after three years in service and had to be replaced. The first replacement was at TMI-1. I was doing fuel tests there at the time. B&W borrowed one of my vibrating tools to help unstick bolts they could not unfasten. They never returned this tool - they still owe me one.

Once all the above designs were approved with comment by Wachter, the actual program of modifying 12 reactors in various stages of manufacture was started (289 million dollars worth of reactor components).

While the failed components were being reworked, Wachter Associates planned the installation of numerous test sensors to provide positive assurance that all design inadequacies had been corrected. This turned out to be the largest test program on a large power generator in the history of the world.

Because of the large numbers of instruments planned for the Oconee I Hot Functional Test Program, this is also a landmark effort in the history of commercial nuclear power plant construction. There are several important features in the program:

1. Utilized weldable strain gages in a high pressure, high temperature water environment;

2. Utilized pressure transducers deep inside the pressure boundary without a drywell;

3. Utilized specially designed weldable conduit covers to shield fragile sensor leads from the dynamic forces of the high velocity water flow;

4. Utilized both split and multi-penetration Conax seal devices to bring sensor cables from surfaces of internal components out through the pressure boundary.

The instrumentation recommendations made by W. J. Wachter & Associates in May for hot functional tests were as follows:

1. Accelerometers - Endevco with 304 stainless steel cases.

2. Strain Gages - Microdot with 304 stainless steel cases.

3. Pressure Transducers - Kaman with 304 stainless steel cases.

4. All leads to be covered by 304 stainless steel conduit welded to the major structurals.

5. Conax seals for passing leads through the pressure barrier.

These recommendations, had they been followed without question, would have greatly reduced the lost time and cost of the Oconee failures. The usual problem of too many experts giving too much advice in a situation like this resulted in unnecessary tests, backup instrumentation, and other costly additions to the program.

The final instrumentation installation consisted of 40 strain gages, 10 accelerometers and 11 pressure transducers. There were 73 leads that had to be routed through the reactor to be attached to these instruments. All the leads had to go through the primary pressure boundary to the array of read-out equipment.

Wachter Associates provided the technical experience obtained from techniques we developed for military and experimental large nuclear reactors such as PWR I and II at Shippingport. B&W engineers had little experience with this type of instrumentation and that little experience was usually wrong. Time and again B&W engineers introduced designs that were inadequate and failed during individual instrument environmental tests.

One major area of contention was the protection of instrument leads exposed to high velocity coolant flow. We recommended encasing all leads in half round flanged conduits that were stitch welded to base components (thermal shield, core barrel, etc.). B&W ignored our recommendation and proceeded with the ridiculous notion you could protect leads using a .015" thick flat sheet of stainless steel tack welded to the major support components. We demanded a flow test be performed. B&W went to their consultant, Chalk River Research Lab in Canada, who recommended this sheet metal shield for leads in the first place. After a short time on flow test all the sheet metal tore off and wrecked the test. As a result B&W went with our design and added strength welds to the conduit instead of stitch welds. We demanded to know how they were going to remove the conduits after the hot functional flow tests were finished. <u>They didn't have a clue.</u>

To verify our determinations on instrument locations on reactor components we requested that vibration tests be performed on a full size model of the Oconee Reactor located at the B&W Research Lab in Alliance, Ohio. Vibrating the reactor proved to be very difficult for the Lab personnel at Alliance. After days of phone discussions, Bernie Langer and I went to the lab and had the vibration tests performed under our direction. All the major components natural frequencies were determined and recorded for use during the final hot functional flow test of the Oconee I reactor.

While the work was proceeding on the reactor internals we turned our attention to the reactor vessel. We had removed the U baffles on the core support shield and redesigned thermal shield, lower grid assembly and elliptical flow distributor, but the reactor vessel still had the flow stabilizing vanes welded to the stainless steel cladding (no strength welds) of the inner vessel. During my inspection of the vessel internal surface I had to climb over these miserable obstructions measuring four feet high at the bottom of the vessel and tapering to a few inches high at the vessel wall. After a long and bitter fight over removing these vanes, I was saved by a mixing vane failure on a primary coolant pump inlet pipe at a Combustion Engineering nuclear power plant. These vanes were strength welded to the pipe. They tore off and destroyed the primary coolant pump. Citing this failure I forced B&W to remove the mixing vanes on the vessel. The next time I inspected the vessel only one inch stubs remained on the vanes.

The installation and calibration of the flow test instrumentation proceeded with some difficulty due mostly to the inexperience of B&W. As the date for hot functional tests approached I began to worry about the major challenge of sealing instrument leads at the pressure boundary of the reactor vessel. The main seal was the Conax multiple split fitting. We performed reliability tests on this seal at the University of Pittsburgh, Mechanical Engineering Lab in August 1972. The fitting was installed in the 1893 pressure vessel that still passed Boiler Code Inspection every year. The system was brought to temperature (550°F) and pressure (3125 PSIG), cycled five times within the prescribed temperature range, then disassembled and examined.

While Wachter Associates were qualifying, the work at the site was coming to a completion and the hot functional test was ready to start. As stated before, my major concern was the primary pressure seal for instrumentation leads exiting the nuclear pressure vessel. A surprise visit to the site was made. B&W had not contacted us about any problems, but I knew they were in trouble. After interrogating them regarding the Conax seals, certain problems emerged. It was determined that several of the Conax fittings were of the wrong type and had to be replaced. Last and most distressing was the fact that B&W personnel were unable to install the fittings. This meant that the hot functional test <u>could not be performed</u>!! B&W ignored instructions and did not make the installation fixture we told them they needed to install the Conax fittings. (We even sent them a set of detailed drawings that showed how to make the fixture.) <u>Wachter had to save Oconee I</u> (plus 11 other reactors) <u>again</u>! I called in the design over the phone to Frank Rhodes at US Tool Co., our fixture supplier. We provided B&W replacement seal components and custom designed fixtures in <u>four days.</u> The instrument ports were sealed and the hot functional tests were performed on schedule.

The hot functional test of the Wachter designed Oconee Reactor provided proof this design met every criteria established by the Langer/Wachter Design Criteria. The only marginal component was the surveillance tube which we wanted to have three point support instead of two point support. This component was strengthened by thickening the tube which added marginally to its strength. This component failed at TMI-1 after two years and it was replaced by our original recommended design.

During the hot functional test a small ticking noise was heard. I was called in to try and discover the source of the noise. I spent eight hours lying under the reactor listening to this noise through special earphones. I determined it was a rattle of the central instrument tube installed in the center of the reactor through the bottom of the reactor vessel. We removed it and put a slight bend in the end of the tube so it had <u>positive</u> contact with the guide tube it was in. A rerun of the test showed the ticking sound was eliminated. The test was completed and all the readouts of the 90+ instruments were reviewed.

The data obtained during the hot functional test confirmed that the Wachter design of Oconee type reactors would perform without major failures and all <u>major</u> components had design safety factors of from four to ten times the strength required of these components. The stresses in the 100 bolts fastening the thermal shield to the bottom flange of the core barrel as measured by four instrumented thermal shield mounting bolts were remarkably low: 2600 pounds per square inch steady stress with a ±1200 pounds per square inch oscillating stress. Since the yield strength of these bolts was 80,000 pounds per square inch it was obvious that the top brake shoe clamp support should be removed. This was never done and after three to five years of reactor operation all these reactors experienced random thermal shield mounting bolt failures due to thermal cycling of the reactors during shutdown and startup. At these times the core barrel can be 5 to 10 degrees hotter than the thermal shield. This difference will cause a 5 million pound force on the bolts which eventually results in random fatigue failure of the bolts.

Upon completion of the hot functional test, removal of test instrumentation was begun in preparation for the actual startup of the reactor. B&W was again silent about a major problem created by their inability to follow Wachter instructions. I received another frantic call from George Kessler, headquarters VP of Engineering. He suspected the "kids" from Lynchburg were in trouble again. I made a trip to the site and inquired about any problems. They finally admitted they couldn't remove the lead protectors they strength welded to the major reactor components. None of the cutting or grinding machines would work. Wachter Associates had discovered a West German grinder that would work. After procuring this tool B&W technicians were trained to use it. This tool later proved invaluable to the field repair work Wachter Associates had to do on other reactors.

After all the instrumentation was removed, the reactor was prepared for startup (nuclear fuel was installed) and a final inspection performed by Wachter. This unit is still generating electricity today after <u>thirty years</u> of operation.

CHAPTER 6
Fort Calhoun Spent Fuel Storage Problems Solved So Reactor Could Start Up and Later Continue to Operate
1973-1974

In the spring of 1973 I received a frantic call from Pickard, Lowe and Garrick (PL&G), primary consultants on reactor systems for Omaha Public Power District (OPPD). During tests of the nuclear fuel storage racks, the dummy fuel assembly used to confirm that the racks would accept the fuel to be stored in them showed bad scratches on its outer surface. The worst scratches were on the zircaloy grids supporting the nuclear fuel rods. This was the first reactor that used zircaloy grids to support the fuel rods. Westinghouse stated that zircaloy grids would fail in the first year. They used Inconel 718 high strength alloy for grids to take the imposed loads supporting the fuel rods, (a wrong assumption based on wrong analyses supported by Bettis - force not stiffness was important). I announced to the utilities that the CE grids would last as long as the Westinghouse, B&W, GE grids - they were all <u>bad designs,</u> but they would work marginally as long as fuel depletion was kept fairly low. The idea of scratched zircaloy grids panicked the utility. We were brought in to find and repair the problem. PL&G told the utility we were the only consultants in the world that might solve this problem and fix it. The racks were welded open frame made up of angles and other structural steel (all stainless). The fuel cells were formed by four angles, one in each corner. Unfortunately every fuel support angle had weld spatter along its full length. The spatter acted like coarse sandpaper and gouged the outer surface of the fuel assembly as it was being inserted into the cell for storage.

The problem was to remove the weld spatter from 10,000 feet of internal angle surface that supported the nuclear fuel during initial loading and later, spent fuel. Normally the fuel would be stored in these racks with 40 feet of water in the spent fuel pool where the racks were installed. During the first loading of the reactor core before startup, new (unirradiated) fuel was temporarily stored dry in these racks and then transferred to the reactor. Without these racks the fuel could not be loaded into the core and the <u>plant could not start up</u>. Delay in startup costs the utility $500,000 a day. With only five weeks before "fuel loading" the utility gave us complete authority to perform this work on a "crash" basis (turnkey project).

It took two weeks to design, build and send the tooling to the site. The first tool created was a long handled tool with a precision grinder at the end similar to the grinder used to remove the instrumentation conduit at Oconee.

The Twentieth Century, The Age of Mediocrity

It had unique control and positioning features to provide a smooth finished surface where weld spatter occurred. The second tool was similar to the first only it inspected the surface. The inspection tool had a special paint roller on the end with plastic over it. The inspector would roll over a section of internal angle and check for indentations on the plastic, indicating weld spatter that was missed during grinding.

After training for five days, two crews performed the actual repair and inspection. Swede Woolard was head of the crews and also worked on the repair operation. Daniels Company supplied some of the technicians. On the second shift they had a mean Indian and they suggested that he and Swede should never work together. One day they ended up on the same shift and everyone held their breath. They got along and became best friends -a Swede and an Indian. The project was completed a week before the nuclear fuel arrived. <u>No one thought we could do the job on time.</u> We packed up and left, little knowing that we would be back next year.

About a year later, close to the first refueling shutdown, we got an emergency call from the utility. They had a storage rack modification that had to be done in five weeks (the shutdown and refueling was scheduled at that time). We were subjected to the usual hassle over completion date of the job with the usual threats of $500,000 a day penalties for being late. After a careful review of the project we told them we could finish the job in time for refueling. This time we had to use utility employees which was a pain. Halfway through the job they threatened to stop work because I was pushing them <u>too hard</u>.

After the first refueling, Fort Calhoun was going to replace the spent fuel racks with high density spent fuel racks. Prior to filling the spent fuel pool, where the racks are located, with water they wanted us to prepare all racks but one for easy underwater removal though not designed to be removed. The remaining rack had to be modified to remain in the pool to take the first spent fuel (one third of the core of 139 fuel assemblies, or 44 fuel assemblies). It had to be modified for ease of ultimate removal (under forty feet of water) and at the same time be anchored to the pool in order to withstand an earthquake. All earthquake (seismic) analyses were done by me. This was another turnkey job. The racks were all tied together with heavy welded bars. The tool developed to remove the instrument conduit strength welded to the Oconee core barrel was modified for cutting the tie bars on the racks at Fort Calhoun. I cut most of these bars myself and trained others to do this job. Once the racks were separated special lifting straps were attached to the racks for lifting them from the pool by an overhead crane. At first we were going to leave the racks in the pool until after the pool was filled with water and the 44 spent fuel assemblies stored in the remaining rack. After a few meetings we convinced the utility to remove all the racks but the one needed to store the spent fuel. This operation was performed without incident. While this was being done, I designed special seismic restraint structures that were rigidly attached to this rack and were attached to anchors in the spent fuel pool walls by sliding keyway attachments that restrained the rack laterally but would allow it to be lifted vertically when the rerack project started. I sent drawings to US Tool & Die and they fabricated 2000 pounds of stainless steel structural restraints to be installed on the fuel storage rack to be left in the pool. The seismic analyses I performed were the first requiring third party review (analyses a lot more complex than the analyses performed by reactor suppliers). I did the analyses at night in my motel room and sent them back to Pittsburgh to one of our consultants who did the third party review. The final document was returned to me and I submitted it to the utility engineers as required by the AEC. The restraints were fabricated and sent via air freight.

The final phase of this project was installing the seismic restraints on the remaining rack. The shipment of this hardware (2000 pounds of it) was by air freight through Chicago to Omaha. The shipment got lost in Chicago and we were approaching the nuclear refueling shutdown of the reactor when the spent fuel would have to be stored in the remaining rack. The Friday before the Monday deadline I chartered a plane to go to Chicago, locate the shipment, and fly it back to Omaha in the dead of winter. I rented the largest station wagon I could find and drove to the air charter terminal at Omaha. We flew to Chicago and taxied to the air freight terminal. I had all the paperwork on this shipment and talked to the people there before coming. The place was closed so we had to break in and look for the shipment. We found a guard and explained the emergency we were in. He helped us look for our missing shipment. After a long time we located the equipment which fortunately was packaged in a dozen separate boxes, none of which weighed over 200 pounds. We loaded the boxes on a wheel dolly and rolled them out to the airplane. The pilot had a hemorrhage when he saw how much stuff there was. Distributing the boxes as the pilot wanted them took another hour and it was 2 a.m. Saturday morning before we took off in a heavy snow storm. The deicing boots on the wings and tail were having a hard time keeping up with the rapidly forming ice. The plane became more and more sluggish and unstable, until we were forced to land at a small airport in Iowa (we almost crashed). After deicing, we continued our flight to Omaha. I drove the station wagon up to the plane and we put the boxes in it. Needless to say, the wagon was bottomed out. I managed to drive it to the utility storehouse outside the fence at the reactor site. We unloaded the boxes and I drove back to the motel. I got about two hours sleep and went to the site. We got the seismic restraints through receiving, inspection and cleaning and finally got to work on fastening the restraints to the rack. Assembling and fitting up these devices took a long time. I was warned about pushing the technicians too hard so we slowed down. After a four hour break we finished the job just hours before the pool was filled and spent fuel moved into the rack. At this point I began to wonder if nuclear power was worth saving.

CHAPTER 7
The Palisades Disaster
Miraculous Recovery by Wachter

Ongoing work associated with the nuclear power industry kept me in touch with the latest developments of methods for monitoring and diagnosing the various dynamic problems associated with the power generating reactors. One technique was monitoring "Neutron Noise" to provide vibratory movements of the reactor. A scientist I knew at Combustion Engineering had developed a method of measuring displacement between the reactor external boundary and the inner wall of the reactor pressure vessel. At this time he was closely monitoring the Palisades Nuclear Reactor. In the summer of 1973 he called me to inform me he was picking up motion of the reactor in the amount of ± 0.010 inches. He asked if this motion could be the elastic deflection of the core barrel in the fundamental "hoop" mode of vibration. I told him "yes." After about three weeks he called to tell me he was measuring a motion of

±0.020 inches. I considered this still to be in the range of elastic "hoop" deflection of the core barrel. The motion became progressively larger through the fall of 1973. When it reached ±0.090 inches, I said, "Shut it down!" Fortunately they did.

After Oconee I became known as the "Red Adair" of the nuclear industry. The evidence I evaluated for Combustion Engineering in October looked <u>very bad</u>. After removal of the head a critical measurement was made. This was the distance from the top of the reactor vessel to the top of the reactor core barrel. This dimension should have been 17 inches. A technician measured 17¼". After checking this measurement a dozen times it was accepted along with the terrible implications of damage to the pressure vessel ledge and core barrel flange. The worst assumption was that ¼ inch of stainless steel cladding had worn off from the vessel ledge. This cladding was less than 5/16 inch thick, leaving very little cladding on the ledge. "Impact fretting" wear was my first thought, which later proved to be right. This type of wear could only occur if the clamping force between the pressure vessel head and the core barrel assembly failed and the core barrel assembly rocked and rolled around the ledge of the vessel causing impact forces to exist between the core barrel flange and the vessel ledge upon which it rested. This would cause impact fretting that would remove metal from the interacting surfaces at an alarming rate. Fortunately neutron noise measurements provide us with the measurement of progressively increasing motion occurring at this interface. This allowed for a timely shutdown before the reactor tore itself apart.

In phase I Wachter Associates Palisades Failure Evaluation Project, as stated by the C.E. contract, was to "Provide professional services for the analyses of the Palisades Reactor vessel internals failures and determine the cause/s of these failures, recommend solutions to the above problems that would allow the reactor to continue to operate." This phase started October 24, 1973 and was completed in four weeks, November 21, 1973. The basic analytical approach was the same as the Oconee failure analysis which is based on the following experience.

In 1968, hydraulic vibratory forces caused major internals failures of two Westinghouse reactors in Europe. These failures could be attributed to overdesign for thermal stresses which led to underdesign for hydraulic load stresses. The same type of failure occurred at Oconee I, combined with failures of long slender bodies projecting into the coolant flow. Again, the analyses indicated no problems, but a review and post test load deflection experiment provided confirmation of hydraulic loads postulated by W. J. Wachter that would result in the above failures. The extension of the analyses performed on thermal shield loading due to hydraulic forces results in a static hold down force of 200,000 pounds required to prevent motion of the core barrel assembly. The dynamic restraint required to stop rocking of the core barrel is 400,000 pounds.

The Palisades Nuclear Reactor core internals were clamped between the head and pressure vessel ledge using thermal interference at temperature as the main restraint. Since the total deflection available is only

0.0105 inch, any small interference during reassembly of the reactor could provide a condition that would permit local yielding of contact points and a progressive wear fretting situation leading to the conditions observed. Failure to properly seat the reactor on the vessel ledge created a clearance that allowed the reactor to lift and the core internals assembly could roll around on the vessel support ledge in a whirling mode of motion similar to the free body motion observed in other reactors. The conditions of the wear surfaces supports the above theory in that a high degree of fretting occurred on bearing surfaces accompanied by rubbing wear on other surfaces. Since the support conditions were highly unstable, the true mode of motion was a constantly changing one made up of several combined motions. At no time were the amplitudes of motion large enough to cause a problem with the actual reactor operation, and the reactor could have continued to operate without any difficulty. As previously stated the Palisades core internals incident is similar to previously observed incidents at Selni, Senna, Obriegham, and Oconee I. Loss of hold down clamp permitted the core internals to lift on one edge and roll around on the edge of the barrel flange. This rolling/rubbing action resulted in fretting and wear at various bearing surfaces. An indepth review of all damaged surfaces show that no major damage occurred and that a minimum of refinishing and repair is needed to place the system back on-line.

The following repair work should be performed:

1. Dress pressure vessel head keyways and seating surface.

2. Leave the upper guide structure as is.

3. Dress the barrel flange as required.

4. Dress the pressure vessel ledge.

5. Replace three sets of snubber blocks.

6. Install a hold down clamp assembly to provide 1,000,000 pound hold down force with a 0.110 inch deflection.

The elimination of clearances in the core internals assembly clamping system, plus the addition of a large accurately measured hold down force insensitive to minor dimensional errors, is the fix required to prevent wear/fretting in the Palisades reactor. <u>The clamping system is the design developed by W. J. Wachter with the assistance of Mr. B. F. Langer.</u> The load member is a cylinder of 304 stainless steel, 138 inches outside diameter by 132 inches inside diameter by 12 inches high. This ring is capable of supplying a clamping load of a million pounds to the core barrel flange. The rotational deflection of the ring will result in a deflection of 0.110 inch at the outer bearing lip of the ring segments. The twelve clamping segments can be placed around the ring in such a manner as to avoid any special instrumentation in the area of the clamp ring assembly. The location and size of the ring can be shifted to further accommodate instrumentation. For some strange reason, based on talking to their German consultants (Seimens) they went with the refrigerator light push buttons (132 of them). These push buttons failed to meet the Langer/Wachter design criteria (only

80% of required deflection and 80% of the required hold down force). The Germans used these things to hold down their reactors at five nuclear power plants.

In addition, certain surfaces should be restored to a condition compatible with the new hold down clamp. The dressing of surfaces can be performed using a semi-automatic grinding system similar to the one provided by W. J. Wachter at Oconee. Information on this equipment was given to C.E. Reactor Engineering.

The final fix was the replacing of three sets of snubber shims on the pressure vessel. This will required the use of local shielding and a diver, or the use of remotely operated semi-automatic tools similar to those used at Selni.

I presented my summary report covering my contract work for C.E. The next thing I knew I had to battle C.E. engineers regarding the question of repairing the reactor. They didn't think it could be done. I had a meeting with the CEO of Combustion Engineering and his technical assistant, Commander Calvert (retired Naval submarine commander under Rickover). To my dismay, Calvert proved to be anti-nuclear power. I talked to John Gibbons on the phone during a break in the meeting and he confirmed this fact. In the end, I prevailed and was given the assignment of presenting my report to the AEC (before NRC existed). At a meeting in Bethesda with the AEC I made the presentation and battled questions from the government for over two hours. I finally convinced everyone at this meeting that the Palisades failure had been properly evaluated and that the fix would be strong enough to prevent future failure. This meeting took place on November 28, 1973.

After a month of arguing about whether Palisades could be repaired, C.E. Nuclear Construction Division declared it couldn't be repaired. This was a "hot" reactor that had to be repaired "hands on." Nobody could do this. A meeting was called at which Wachter Associates declared the Palisades Plant could be repaired and be in operation by early spring (the meeting was in January 1974). At this time the manager of C.E. Nuclear Construction told me, "If you're so smart, come and fix it!"

The Phase II of the Palisades Project covering the repair of the reactor was issued the end of January 1974. This contract stated, "The Seller, Wachter Associates, shall furnish labor, tooling, training, procedure guidelines and consultation in connection with work done in Wachter's shop and at the jobsite of Consumers Power Co., Palisades Station in South Haven, Michigan.
 The work will include rework of the keyways in the nuclear pressure vessel, rework of the core support ledge surfaces, rework of keyways in core support barrel flange for pressfit of keys, reconditioning the edge of the underside bearing surface of the core support barrel if necessary, and rework of keyways in the head."

Two members of Wachter & Associates (S. Kaufman and J. A. Weissburg) met with members of Combustion Engineering at the Windsor plant to discuss the capabilities of Wachter & Associates in the area of field service work. Next, a visit was made to the Palisades Plant by Mr. Kaufman and Mr. Weissburg, with a preliminary view taken of the reactor area. A review of the problem and the experience of Wachter & Associates was held with Mr. Gordon Dick, C.E. Nuclear Construction Superintendent, at the Palisades site. At this time, it was agreed that Wachter & Associates would attempt to have personnel and some of the repair equipment at the Palisades Plant by January 7, 1974. An additional meeting took place at Windsor, with an outline of the proposed repair methods presented, and further information supplied by C.E. concerning the extent of damage determined to date. Rubber casts of some of the damaged surfaces were examined.

In summary, the damaged areas which were to be of concern during the repair program developed by Wachter & Associates were identified as follows:

1. Local raised spots (1 inch diameter or less, 0.03 inch high or less) on the worn horizontal support ledge of the reactor pressure vessel.

2. Wear-displaced (0.08 or less) lateral surfaces in four reactor vessel keyway cavities. Some worn surfaces were smooth and parallel to the original surfaces; others were rough, bumpy or stepped.

3. Lower perimeter corner of outside diameter of reactor core barrel flange. Since the barrel flange had worn a form-fitting cavity into the reactor vessel ledge during steady-state operation, (0.23 inch maximum), it was considered necessary to remove a portion of the outside of the core barrel flange to

accommodate the lower position (0.28 inch maximum) of the core barrel during startup transients when the expansion of the hot core barrel flange may not yet have been matched by the cooler reactor vessel ledge region. The amount of material to be removed was to be determined later.

4. Wear-displaced lateral surfaces of four core barrel flange key retaining slots. These slots engage in an interference fit with the alignment keys and provide the surfaces which set the position and attitude of each of the keys in the reactor assembly.

5. Underside horizontal worn surfaces of the core barrel support flange (0.069 inch maximum) resulting from wear against the vessel ledge. Some metal to be removed to accommodate transient expansion as in (3) above. The amount of material to be removed was to be determined later.

6. Wear-displaced lateral surfaces in four reactor closure head keyway cavities. Some worn surfaces were smooth and parallel to the original surfaces; others were rough, ridge-marked, striated, or stepped.

It was decided that Wachter & Associates would design, develop, procure, deliver, provide training for, and technically supervise the field operation of corrective equipment to be used on the reactor components at the Palisades Plant. It was also agreed that Wachter personnel would be consulted as to the degree of rework necessary and advisable, as well as the establishment of inspection acceptance criteria for the rework, based on Wachter & Associates reactor design and analytical experience. In some cases, the interrelation of some of the damaged areas had not yet been fully evaluated, and it had not yet been decided how many of the surfaces would be refinished, and how many could be accepted with minor or no refinishing. One reason for caution in refinishing was to assure that the remaining cladding, (only 0.030" in some places), would not be penetrated during refinishing of any of the severely worn areas.

Because of the problems associated with the local rework of irradiated reactor components in the plant reactor compartment, it was recommended by Wachter & Associates and approved by C.E. that small, individually-fixtured local machine setups be used to rework each keyway area, less complex than a typical shop machine tool setup, but more precise and freer of operator dependence than purely hand-operated cutting or grinding equipment. Mock-ups were to be used extensively for tests of tooling and techniques, as well as for operator and quality control training. When considered appropriate, procedures were written with provisions for previously approved optional methods, so that the cognizant engineer would have the freedom to select the best method for quickly, safely, and accurately achieving the desired results at each specific work location. It was realized that conditions such as radiation level, material hardness, surface finish, condition of nearby reference surfaces, obstructions caused by immovable fixtures and equipment, and other factors not present on the mock-up might cause one method to be much more satisfactory than another, though they seem equal at first trial.

Both abrasive grinding and multitooth-cutter machining were employed for the power refinishing work. Hand filing was used for rounding of local edges and removal of burrs. Grinding was used to generate the precise flat keyway faces. A three inch bonded carborundum cup wheel in a right-angle grinder was used for all three types of keyways. A half-inch solid tungsten-carbide side-cutting rounded tip milling cutter in a long-shank straight drive unit was used to remove all the local raised spots from the vessel support ledge. The carbide cutter was also used to remove some of the back-radius steps left on the larger diameter of the abrasive wheel. The abrasive unit was mechanically guided by a clamped metal fixture for each application. The carbide cutting unit was carefully hand-held for its various applications. Tooling was made of stainless steel or heavy aluminum plate coated with a polyethylene strip or inert synthetic lacquer. All precise reference joints were machined, doweled, and bolted, not welded. Tools, fasteners and equipment which might otherwise have fallen into inaccessible areas of the reactor were secured with lanyard cords. Shatterproof acrylic plastic mirrors were used to reduce setup time while minimizing personnel exposure to vertical streaming radiation between the core barrel O.D. and reactor vessel I.D. Mirrors were used extensively during mock-up training, and confidence in their use was soon developed by all the craftsmen. Similarly, the mirrors proved quite useful when working under the reactor closure head. High resolution, low magnification binoculars were used to conduct preliminary visual examinations while reducing actual radiation exposure. Plastic wrapping was used to protect the binoculars from glove-borne contamination. Similarly, micrometers were pre-wrapped on all knurled surfaces to prevent unnecessary contamination.

An open-sided shielding support box was built to provide a radiation shielded shell in which two men were sheltered during measurement and resurfacing of closure head keyways.

It was determined that preliminary, intermediate, and final measurements be taken of each keyway cavity, and that two independent, qualified agencies provide the measurements. In each case, Wachter & Associates served as one of the agencies. In almost all others, (C.E. Chattanooga personnel made preliminary head keyway measurements [W. Milligan] and final measurements were made by Wachter & Assoc.), P. F. Avery personnel made the same measurements. In all cases, measurements agreed to within 0.008 inch, with the exception of those taken across extremely round or irregular faces. Measurements were taken either with snap gages or inside micrometers used as snap gages, the final readings being taken by calibrated and certified outside micrometers reading the length of the inside gages.

On January 7, 1974 we arrived at Palisades with tools and equipment ready to start work. Upon examining the main floor of the containment building we determined it to be unsafe. The floor was littered with air hoses, electric lines, trash, etc. We spent three days cleaning

The Twentieth Century, The Age of Mediocrity

up the area that looked like the German retreat from Moscow (evidence of how traumatized the Palisades people were). The first tool setup was started, but immediately ran into difficulty. The 1", 8 threads per inch mounting bolt holes in the top of the pressure vessel were left hand threads instead of right hand threads. <u>No one at this site knew this</u>. I called Frank Rhodes at U.S. Tool, our supplier of special equipment. He made eight left hand thread bolts and shipped them by Greyhound Bus (the original FedEx) that night.

We built mock-ups of the various repair locations to train personnel in the actual performance of the work. This area was outside the reactor containment building and it was very cold. Swede Woolard was head of the actual repair operations. He came from Bettis where he was a legend in his own time. He assembled the first nuclear reactor as well as all the ones that followed - an expert machinist, mechanic, welder, etc. During the height of the repair work I received a visit from the head of the union. He said this was not a labor dispute, but his people were concerned that Swede was working too hard. He said, "If anything happened to Swede (he was over 70), we are all dead. He's the only one who knows how to do the work." Later I found out that the technicians who worked for him were feeding him and providing a rest area for him in their trailer. The workmen were from Bechtel. We were assigned a trailer next to the containment building that was abandoned. The reason it was empty was because it <u>smelled bad</u>. We never found the cause of the smell, we just got used to it. Prior to starting the actual repair, I met with the QA manager whose name was Silas

Heath. He was bad news from the beginning. Because of lack of confidence he overdid everything and was a real obstacle to our progress. Another problem developed over the crews we trained to do our work. Often these crews were taken from us to work on the steam generator repairs in progress at the same time as our repair project. The C.E. people would use up their allowed irradiation exposure and they could not return to our work. Nevertheless our work proceeded to our schedule of ten weeks to complete this repair.

Based on the evaluation of data, certain keyway surfaces were designated for complete restoration, others were designated for local spot removal and blending, and others were determined to require no significant surface correction except for corner deburring. Work was first performed on reactor vessel keyways and vessel support ledge raised spots. This was done with the core barrel in the vessel, but supported above the ledge by three 20" high blocks. Ledge spots directly under the blocks were removed later. After the preliminary vessel work was done, the pit was flooded, the core barrel was raised out of the vessel and moved to one side of the pit. The upper three-foot portion of the barrel was out of the water. At this time, the barrel keyways were ground to provide proper surfaces for the 0.002 to 0.005 inch interference fit with the alignment keys. Workmen were supported on temporary wood and metal platforms for this work. In this same general time period, but during alternate (day) work shifts, divers performed underwater repair and replacement of snubber shims at the bottom of the reactor vessel. The concept of using divers for reactor repair work was a result of my seeing pictures of Italian divers swimming in the primary coolant pipes at Selni in <u>wet suits</u>. If they could get away with that I decided we could use hard hat divers. These men came from working in the ocean with zero visibility. They thought they were in heaven working in the reactor vessel. They trained in a clean pool at the C.E. plant in Windsor, Connecticut. Any leakage of the standard deep sea diving suit was unacceptable, so Wachter Associates redesigned the suits to have zero leakage. The hard part of their operation was drilling locking pins out of 24 bolts holding the shims. It took them fifteen to twenty minutes of hard hand drilling to remove each pin. From Sears we got a Shop Smith that was modified to be a portable drilling machine. Using this tool the divers were able to drill a pin in less than a minute. All twenty four pins were removed in one dive.

Work was initiated on closure head keyways using fixtures modified from those used on the vessel keyways. Meanwhile, the barrel was returned to the vessel with the support blocks in new circumferential locations. The pool was drained and the few remaining spots on the vessel ledge previously obscured by the support blocks were machined away. Work on the closure head and vessel flange was completed during the same shift. Tools and fixtures were removed, checked, decontaminated, and returned to normal work areas for future use. All grinding units experienced internal contamination and were retained in secure areas awaiting storage.

Reactor vessel keyways were repaired as follows:

$270°$ keyway, $180°$ side, complete resurfacing

90° keyway, 0° side, spot resurfacing

0° keyway, 90° side, spot resurfacing

180° keyway, 90° site, spot resurfacing

Core barrel keyways were repaired as follows:

0° keyway, both sides, complete resurfacing

180° keyway, both sides, complete resurfacing

Closure head keyways were repaired as follows:

180° keyway, 90° side, complete resurfacing

For all keyway resurfacing, a surface finish of 32 to 16 micro-inches was achieved. This is smoother than the finish of 64 micro-inches required by the design specifications. Approximately 60 spots of various sizes and heights were removed to within 0.005 inch of the nominal local ledge surface of the reactor vessel.

Corner removal, 0.25 inch by 0.25 inch, 44 feet of circumference; underside removal, 0.05 inch thick by 0.75 inch wide, 42 feet of circumference of the core support barrel flange. In all cases, work progressed smoothly and steadily after each stage was described in approved procedures and officially initiated. The success of each stage can be attributed to a combination of the following:

1. Development of logical procedures, work methods, and inspection techniques and acceptance criteria; adherence to procedures.

2. Careful training of craftsmen, foremen, inspectors, and engineering personnel on mock-ups.

3. Extreme familiarity of craftsmen with tooling, fixtures, and work methods.

4. Dependable performance of tooling and fixtures, as an aid and supplement, not as a replacement, for the skill of the craftsmen.

5. Excellent and timely support from various on-site service groups, such as carpentry and machine shop.

6. Cooperation and continuous support from C.E. Nuclear Construction, C.E. Reactor Engineering, P. F. Avery, and Chattanooga personnel.

7. Continuity of technical supervision by Wachter & Associates personnel.

8. Procurement support from Wachter & Associates off-site personnel.

Upon completion of all the repair work a visual (underwater camera) showed a sinister dark area on the lower side of one of the core support barrel outlet face seals. It appeared the lower edge had been worn away thus making the component unreparable. All work was stopped for three days and the project appeared to be doomed. A disaster for the whole nuclear power industry. On the morning of the fourth day I walked into Mr. Flynn's office and saw a C.E. calendar on the wall. The month of February 1974 had a picture of the above unit mounted on the vertical boring mill. The same dark area was present in the picture. <u>It was not a damage area.</u> I immediately restarted the project. Final inspections were performed on all components except the pressure vessel head keyways. At this time there were no project people who hadn't acquired their allowable dose of irradiation. The only people left were Mike Milligan, C.E. Field Superintendent, and me. We had to wear special masks because of beta rays coming off the underside of the head. Betas penetrate and damage eyes if not protected by 0.2" thick plastic. The technicians (Vallee's) removed our glasses and put the special hoods over our heads. Mike said, "If you take my glasses I can't see to read the micrometer!" I chimed in, "That goes for me too!" So they built us special masks that would allow us to wear glasses. We crawled under the head and made the keyway final measurements. After that Mike and I shook hands and said goodbye, little realizing I would encounter him at the FFTF site years later. The reactor was reassembled and started up. It has been operating successfully for the last 29 years without failures. (Good fix!)

In March of 1974 a report on the field rework results of the reactor components at the Palisades plant was issued to Combustion Engineering.

CHAPTER 8
Loss of Flow Test Reactor Project
1973-1976

In 1973 I was called on to participate in the design and construction of a 3/4 scale nuclear power plant to be built at the Idaho Nuclear Reactor Test Facility. The project prime contractor was Aerojet Nuclear Company. They retained MPR as the consultants to design the reactor. The project manager (a former Bettis employee) wanted me to be his overall consultant in the design of the reactor with a special concentration on the nuclear fuel, since he considered me to be the foremost expert in this area. This nuclear power plant was intended to test the nuclear fuel during a "loss of flow accident." The plant was named LOFT (Loss of Flow Test). Since I was already working on Oconee and later Palisades, Aerojet wanted me to be in direct communication with them at all times and be available for meetings. In order to facilitate communication the government gave me two FTS (Federal Telephone System) direct lines to

Idaho National Lab. The government agency directly involved under the AEC was ERDA (Energy Research Development Administration). Design review meetings were held monthly (often twice monthly) in early 1973. The reactor design was based on the S5W concept with fewer modules. Major support at temperature in this concept was gained by thermal differential expansion of major components. The clearance per component (module) was less than 0.010" in the design which was sufficient for more than 20 modules but too small for the few modules in LOFT. Since the fuel had to be removed several times during testing, I told Aerojet they would never get the reactor apart with only 0.010" clearance per module. We changed the clearance to 0.030" which proved to be adequate for easy removal of modules. In general, the nuclear fuel assemblies represented the commercial fuel in use at nuclear electric power plants. (No better, no worse, but not very good.) The important point I made was that this loss of flow test should demonstrate how the average commercial nuclear fuel would behave under these adverse conditions (temperature transients of over 1000°F). In the ongoing construction period of LOFT I made several visits to the site. The site was the old aircraft nuclear reactor site. The construction laydown area was in the world's biggest aircraft hangar (never used). The pallets on which construction material and plant components were placed had strange wiggly lines in the dirt going back under them. We were told these tracks were rattlesnake tracks left by snakes hibernating for the winter. This didn't make us or the workmen very happy. The overall construction of LOFT progressed on schedule with little comment other than correcting errors in the reactor design. Review of the nuclear fuel continued with visits to Exxon Nuclear in Washington State. I commented on fuel pellet chips in the assembly line and other problems that were corrected. My concern at the end of the early fuel fabrication phase was that the fuel would be too good and not representative of the commercial fuel.

At this time Wachter Associates embarked on the main area of our contract. We were to provide inspection systems, inspection programs and training in the use of the various inspection tools we would provide.

The inspection program was developed in two phases. Phase I was the in situ inspection of fuel modules with no removal from reactor vessel. It consisted of three modes of inspection, all to be performed in a minimum amount of time so as not to delay the overall reactor test schedule. The three modes of inspection were QT1 (Quick Turnaround) done with the reactor head in place; QT2 which was performed after removing the head and installing a small inspection tank allowing the raising of the central module for close examination, (note, the central module experienced the most severe service); and QT3 which involved a larger tank and the ability to raise all the fuel modules. The inspection equipment was designed using commercially available components.

The tests proposed in QT1 required no additional equipment, although instrument calibration methods and sensitivity might have to be refined. All instrumentation systems for QT2 and QT3 examinations were designed to use only standard hardware available to the industry in its most economic configuration yet providing a maximum amount of data directly related to loss-of-coolant accident.

The following conclusions were the result of this review:

1. All interim examinations of LOFT fuel modules could be performed in the containment building.

2. Most of the information about core condition between tests could be provided by QT1 inspections prior to opening the reactor.

3. Detailed examination of the central module could be performed in the QT2 inspection tank assembly.

4. It was possible to examine every module, including instrumentation, in sufficient detail in the QT3 inspection tank to make the decision to reuse it or send it to final examination.

The decision to remove the fuel module from reactor testing led to the Phase II inspection, consisting of complete nondestructive and destructive inspections of the module. The primary objective of this program was to determine the past history of the temperature excursions experienced by the fuel during loss of flow tests through examination of the resulting metallurgical and physical characteristics of the cladding and, finally, the fuel pellets. Hydride distribution in the cladding would give a good indication of temperature gradients and local hot spots, particularly in the area of a clad failure. Another equally important objective of Phase II was the determination of structural loads imposed on the fuel assembly and the complete geometric description of the resulting fuel rod coolant channel configuration. In order of importance, inspections would be performed to fully define the condition of the upper and lower tie plates (where currently failures were occurring in BWRs), support tube, instrumentation and control rod assembly, if present. The following section presents the necessary inspections and defines what can be determined by each inspection. The equipment used would be available at TAN-600.

A major consideration in the Phase II program would be the disassembly of the module in a manner least destructive to the module assembly. Planning and care in this operation would be of the utmost importance in order to secure a maximum of useful information from the inspection operations that would follow this disassembly. After reactor testing, fuel rods would be particularly susceptible to handling damage, and special techniques would have to be developed to carefully remove rods from grid assemblies.

The current plan would be to move the module into the vestibule pool and, using the equipment built for Phase I inspection, perform an in-depth reinspection. Color photographs would be made of cladding in the hottest areas for evaluation and comparison with the control pictures previously used in Phase I.

The inspection program in Phase II would be limited to examinations directly related to the LOCI tests and the data required to accurately analyze the effectiveness of the ECCS systems of a particular reactor. At the completion of Phase II, the various parts of the modules would be stored for future examination. The information available on other aspects of the fuel and structural components of the module could be obtained when desired.

If a module failed the QT1, QT2 and QT3 examinations it would be removed from the LOFT containment building. The Phase II inspection would be conducted on modules after removal from the LOFT containment building. During the period when the module would be too hot to be handled in air, non-destructive examination would be performed in the vestibule pool. These examinations would consist of geometric measurements, visual and ultrasonic examination of external fuel rods with special attention to features of the rods that reveal extreme temperature excursions during testing. After underwater examinations, the module would be severed at the tie plate to support tube joint and the two pieces moved to the hot cell area.

The hot cell examination of the fuel module would be performed as described in Item 4.0. The results of these inspections would provide the following results:

1. Measurements defining the peak clad temperature of the hottest fuel rod during the LOFT tests.

2. Physical condition of the cladding.

3. Measurement of peak fuel pellet temperature and thermal gradient through the pellet.

4. Approximation of buckling loads and other conditions revealed by geometric changes in rods, grids and tie plates.

5. Evidence of flow blockage as indicated by "hot spots" and hot regions in the same geometric location of several fuel rods.

This project was the longest in terms of time. It lasted three years. Most of the time was spent developing the inspection systems. One of the areas requiring extensive testing was the lighting system for the optical examination. Some of this work actually took place in a swimming pool. The test program confirmed early fear that the fuel assemblies were manufactured to a higher quality than the assemblies produced commercially. As a result, no fuel rod failures occurred during the actual test program although fuel temperatures exceeded $1000^\circ F$. Before the final operate-til-failure test could be performed, Three-Mile Island accident occurred and the LOFT tests were terminated as having lost their meaning.

CHAPTER 9
DC Cook Ice Condenser Ice Loss Problem Solved - 1975

One of the more ridiculous inventions that had to be handled was the Ice Condenser System intended to reduce the size and therefore the cost of the nuclear reactor containment building. During a loss of coolant accident the hot fluid from the accident heats the air in the building thus raising its temperature and pressure. The size of the cylindrical containment building is determined by the volume of air in the building and how hot it becomes during an accident. In the building with the ice condenser, the hot gases from the accident flow over ice and are cooled so that the pressure is lowered for a given volume of air and the total volume of air can be reduced thus reducing the size of the containment building. There are two problems with this invention. First, the size of the containment building is already too small for ease of maintenance of reactor systems in containment; and second, maintaining the ice over the life of the plant is a monstrous job. The immediate problem that had baffled everyone from the prime contractor (Westinghouse), to the main consultants Pickard, Lowe and Garrick (PL&G), was the fact that the ice was subliming off at an alarming rate.

After several months of effort by Westinghouse, overseen by PL&G, it was obvious they weren't getting anywhere. PL&G called me in to save their tails, as usual. Most of the effort was devoted to measuring the ice contained in over two hundred 14 inch by 20 foot tall ice baskets. Westinghouse claimed to be developing exotic equipment at Waltz Mills. It was estimated the job would take several months. When we were called in, we visited Waltz Mills and got a key to the place where the research was being conducted. There was no one around and the calendar on the wall was two months out of date.

At this point, Wachter decided to go his own way. Jim Weissburg and Sid Kaufman of Wachter Associates took on the ice measuring problem while I worked on the ice loss problem. Jim devised a long handled tool similar in some ways to the long handled tool he created to grind the inside of the nuclear fuel storage racks at Fort Calhoun. This tool consisted of a telescoping pole with a probe type measuring device at its end. The probe measured the distance from the ice basket wall to the surface of the ice. The measurement was read using low power binoculars. Readings were taken at four points around the basket every two feet up the 20 foot tall basket. Knowing these measurements the actual volume of ice in each basket could be calculated.

The Wachter tool was designed and made in less than a week. Using a crew made up of their employees and an ice expert from the Military Polar Lab in Maine, we measured all the ice in one twelve hour shift. While the ice measurement was going on I spent time examining the overall system. The ice was housed in a large annular compartment on the wall of the containment building sealed at both ends by pressure operated doors. If a loss of coolant accident occurred, doors on the inlet side of the compartment swung open and hot gases from the accident flowed over the ice baskets and through pressure operated outlet doors thus cooling the gases. The problem appeared to be leakage around the doors. I examined the seals around these doors. They were worse than the seals I had on my garage door at home (completely inadequate). Leakage of warm air through the ice compartment caused the ice to sublime off at a high rate. New seals were designed and installed and the ice loss was reduced to an acceptable level.

It was interesting to note that PL&G took our information and tools and repeated this job for Duke Power - never giving credit to Wachter Associates. This happened on every project we worked on. The high priced consultants, when they weren't selling "snake oil" relied on us to save their tails. The NSSS suppliers were the worst at doing this.

CHAPTER 10
Fast Flux Test Facility Design Review and Field Consulting Service Including Furnishing Emergency Equipment and Hands On Help to Get Job Done - Saved Project, 1974-1975

Early in the 1970's I received a small consulting contract from Nick Petrich, a manager in the newly acquired FFTF Project. This contract was a design review of the reactor core. The Fast Flux Test Facility was the principal facility planned by AEC for development of sodium fast breeder reactors (ultimately the key to eliminating storage of long lived radioactive waste as well as providing unlimited energy for the 21st and 22nd centuries). The original core had a skewed configuration. This came about during the second and idiotic phase of the Naval Nuclear Propulsion Program. Rickover told congress that the reactors that powered his nuclear fleet could ultimately be refueled as easily as changing a light bulb. (Nonsense!) This led to the idea of refueling by unit module through the head without removing the head. During a visit to GE KAPL labs, K. V. Smith was asked "How can you refuel through the head?" He took a bundle of pencils (7) and twisted them. The pencils separated at the top so that they could be pulled through the head while maintaining ligaments between holes. (You should perform this experiment with pencils to get a hands-on feel for the skew.) This led to huge mushroom heads and fuel modules slanting in all directions. The complication of the skewed core cost the taxpayers millions. The end result being, refueling through the head was <u>slower</u> than removing the head. Someone came up with the brilliant idea of returning the core geometry back to where it was - <u>cylindrical</u>.

Having been in the battle to kill the skewed naval cores, I was faced with killing the FFTF skewed core dreamed up by Battelle Northwest Laboratory (always marginal in their technical capabilities). Battelle Columbus Laboratory had better engineers than Battelle Northwest and provided more valuable contributions to the nuclear industry.

Westinghouse convinced the AEC to change the geometry of the FFTF core from skew to cylindrical and we all breathed a sigh of relief. This reactor with all the test loops to be incorporated into its design was going to be complicated enough.

John Taylor became vice president and general manager of the FFTF Project. I spent many lunch hours in his office discussing the fabrication and site construction phases of this project. When the construction phase of the project had reached the critical stage of installation and assembling the reactor system, I received a contract to be the primary consultant overseeing the work. Fred Klingener, a <u>brilliant</u> and <u>creative</u> engineer was my site manager. If his advice and insight had been used on the front end of the various operations instead of after the fact review, millions in cost and months in completion time would have been saved. We worked for the Westinghouse representative at the site and we had an office in the WARD group.

Initially Fred and I went to meetings on the work ahead. It was obvious after reviewing the site drawings that this system was going to be one of the most complex facilities of its kind. The need for advanced pre-planning became our battle cry. We strongly recommended full scale models of the piping layouts (similar to what was done in the submarine program). The initial work on site construction suffered from lack of pre-planning and scale models. Piping systems kept intersecting other piping systems. Pipes kept trying to occupy the same space as other pipes. Gradually Bechtel supported our model proposals and once initiated they did a great job with it. Later, on the SNUPPS projects (Standard Nuclear Utility Power Plant System) they proudly showed me a building full of models of every system involved in the SNUPPS plants. (I was a consultant to the overseer of this program.)

The other problem causing major delays in the site schedule was the lack of precise detailed procedures on the work to be done. This project relied on the experience and skills of the workers to get the job done. Although the job did get done, it took two to three times as long as it should. Fred Klingener produced procedures for the more complex field operations that were followed after all else failed. Records of the errors and failures with corrections were produced by Fred Klingener and became part of the basis for Wachter Associate's report to John Taylor, "Constructibility Guidelines and Criteria for Designers of Sodium Fast Breeder Reactors and All Other Nuclear Power Systems."

It is impossible to cover all the various installation of components in this chapter. However two major installations will be addressed as well as field equipment problems.

The crane hoist system caused Fred Klingener a great deal of worry. It was a two cable double drum system furnished by one of the most experienced companies in furnishing and operating large crane hoists for field construction sites. Klingener wanted them to wind the cables on the double drum counter to each other, winding on the left drum starts on left edge of the drum and winding on the right drum starts on the right edge of its drum. This counteracts and balances the side forces on the cables so they cancel out. The crane operator had wound the left drum from its left edge and right drum from its left edge. This caused the lift to drift to the left as it was lowered. Initial lifts were a fiasco until Klingener stepped in and calculated how much the component being lowered would drift from the original center as it was lowered. The large components drifted off center as much as five to six inches.

The Twentieth Century, The Age of Mediocrity

There seems to be little support for the procedure process among program management, although to my knowledge, no program has been defined to the extent that merits and shortcomings may be debated. Bechtel Control Document 200 comes closest to defining the responsibilities of each party, but there are few guidelines as to what should be the content of the procedure itself and <u>no guidelines to the reviewers</u>. The result has been the analog of law practice without precedent or constitution. Each individual works within their precepts. The result has been requirements that look like poor procedures, procedures prepared for reviewers instead of the craftsmen and review comments shaken out of enormous nit-nets rather than removed from substance-spears.

The lack of sympathy for the procedure system is further evident in the form of the management hierarchy of both Bechtel and HEDL. Because preparation of procedures is regarded as a peripheral nuisance, Bechtel has a separate "procedures group" rather than technical writers assigned to installation cog engineers. HEDL has a separate "installation methods" review coordination group with neither responsibility nor authority to assure that procedures are faithfully executed. In fact as responsibilities are now constituted, the installation methods group is not permitted to correspond formally with Bechtel either administratively or in the field. This, in particular, promises to be a critical bottleneck in processing PCN's.

Since many of the program deficiencies seem to arise through a lack of definition of goals, terms, responsibility and authority, the cure appears obvious. DEFINE!

It should be noted that the distinguishing features of the liquid metal plant relating to site assembly is the size and relative fragility of the thin wall tanks and vessels which make up the reactor and heat transport systems. The care required in handling and installing these components constitutes a departure from the experience gained in BWR and PWR plants. In this regard, it is important to recognize that typical site rigging is fairly crude by shop or shipyard standards, functionally deficient in precision, stability and mobility. In a real sense, systems must be devised in some cases to protect the component from the rigging. This particularly true where sets are made in tight clearances. The "tightness" of a fit is conveniently defined in relationship to the ignorance factor due to rigging friction. (For example, a component lowered using a simple dowel pin where clearances are on the order of pin radius times coefficient of friction, is in a tight fit.) Isolation and guidance devices should be designed as a system, with guidance in each of the six degrees of freedom as orthogonal as possible, that is, operation or adjustment in one dimension should affect orientation in other dimensions as little as possible.

Certain heavy lift rigging configurations are capable of exerting considerable loads on hanging components. An example of one of those is the double sheave block system used on the FFTF reactor vessel. Unless the lift pin runs perpendicular to the spreader beam, sheave frictional loads may be transmitted directly to the component. A suitable isolation device would be a spherical joint or a double clevis with perpendicular lift pins between spreader beam and vessel lifting device.

As mentioned before, guidance systems should provide control in each free dimension with adequate stiffness. Depending on relative dimensions and weights, hanging components can exhibit enormous rotational stiffnesses about the horizontal axes and special devices may be necessary in addition to those provided for location in the horizontal plane. Many concepts may be exploited for this purpose.

One of the most complicated components of the FFTF reactor was the pressure vessel head assembly. The head itself was 24" thick and 20' in diameter with a complex assembly of stainless steel sheet metal arranged in layers to reflect the heat rising from the pool of 1000°F plus sodium. This assembly was held together with mechanical fasteners. I was concerned that there may be loose parts in this thermal shield assembly. This component was fabricated by Combustion Engineering. (They knew me from Palisades.) Receipt inspection discovered loose parts inside the thermal shield assembly attached to the head. Combustion brought in a special team to fix this problem. My old friend Mike Mulligan was in charge. He and I did the final inspection of the Palisades reactor vessel head keyways.

Special tools were made to perform the reassembly operations and the work progressed at a reasonable speed. Mike and I had dinner at the Hanford House upon completion of the repair on the head assembly.

The next morning I was waiting at the entrance to the clean area where the head was awaiting removal to the reactor building to be installed on the reactor vessel. Mike came out and scowled at me, then said, "O.K., come on in and inspect the repair work." After being dressed in a special suit designed to keep my dirt from the clean head, I went in and examined the job using a borescope (fiber optic viewing device) and special binoculars. That's the last time I saw Mike.

The most important contribution we made to the success of the assembly of the FFTF reactor system was during the installation of the reactor vessel head. The problem facing the Bechtel installation crew was the following: fit a 20 foot diameter head into a 20 foot 0.040" hole (0.020" radial clearance). The depth of the hole was 20". There was no lead in, the bottom edge of the head was a sharp corner. As they lowered the head the foreman looked to me for help. I counted the people standing around (16) and told him to get 16 shoehorns made of 0.06" thick stainless steel sheet curved to fit the holes. Using the shoehorns we got the head started into the hole. Gradually the crew lowered the head into the RV. After ten hours of creeping the head into the hole it finally stuck with 5" still to go (in 15"). After over two more hours of frustrating attempts to seat the head the Bechtel manager turned to us and said "You're the consultants, do something." Fred jumped up on the head and made a small adjustment to a leveling device he had placed on the rig used to lower the head. There was a "chunk" and the head was seated. At this point the Bechtel manager turned to us and said, "I don't know how much Westinghouse has paid you, but you just earned it all in the last ten seconds!!"

The assembly of the FFTF systems proceeded without incident. Upon completion of assembly operations, start-up tests and heat-up tests of the reactor were started. We were asked to calculate the time it would take to heat up the reactor vessel head during the reactor initial hot tests. Westinghouse set up computer analyses using their biggest heat transfer computer program. Fred Klingener analyzed the problem and came up with a closed solution that did not require a computer. The Westinghouse analysis resulted in a time of <u>three</u> days to heat up this component. Fred's answer was <u>17</u> hours. The time required was <u>19</u> hours. This proves that a good analyst can beat a computer jockey any day of the week.

Once the FFTF reactor became operational we closed our office at the site. At this time we were contracted to provide a summary report on installation problems at FFTF. As a result of his excellent work Fred received a commendation from Westinghouse.

The next project we worked on was the Clinch River Breeder Reactor. Fred Klingener was a prime consultant on core design and installation problems to be faced in this program. Unfortunately the project was cancelled by Jimmy Carter. We were assigned the task of writing a document entitled "Constructibility Guidelines and Criteria for Design of Sodium Fast Breeder Reactors."

CHAPTER 11
Spent Fuel Problem, 1976-1978

After saving the power end of nuclear power, a more ominous problem (politically only) reared its ugly head - "storage of spent fuel". This problem was created by President Carter when he stopped nuclear fuel reprocessing and the fast breeder reactor. In our original concept of unlimited nuclear energy, step one was design and build the simplest form of nuclear power generators - light water reactors. The spent fuel from these reactors would be reprocessed and the "hot" stuff would be used to fuel fast breeder reactors. These reactors could use all the useless "depleted uranium" generated from the "bomb days." (This represents more energy than coal, oil and gas combined. In fifty years we are going to need all the energy available in the world.) The energy created in this process would be endless in amount and length of availability. This seemed and still does seem, the only answer to the <u>endless</u> population growth in the world over the next 100 years. Nuclear power uses little land area (not like wind and solar), and creates little waste (compared to coal plants), especially if you burn up all the nasty waste that is "hot" for thousands of years.

Unfortunately the communists created the anti-nuke movement in Europe. They told me at the end of the "cold war" when some of them were my house guests, that the U.S. was the only place they didn't have to create the anti-nuke movement. They had enough "friends" here to carry out the American anti-nuke movement.

The problems created by the liberal left made storage of spent fuel a major political problem. I had already recognized this problem and invented storage systems that would increase storage capacity by four to one and later increase capacity to six to one. These inventions were infringed on by all the NSS suppliers and other bandits like Holtec. (240 patents currently reference my patent as basis for their concept.) This group of inventions saved nuclear power in the short term and saved the nuclear industry, including utilities, billions of dollars in unnecessary dry storage.

Over the next few years Wachter Associates dominated the spent fuel rack industry fabricating over 30,000 spent fuel storage positions installed in spent fuel storage pools at reactor sites (still the safest way to store spent fuel). The business grew so fast we ran out of capital (second mortgage on my house) and had to sell out. We inventors always go broke unless we have good business managers like Edison and Westinghouse had.

The design development and analytical effort to provide the strongest, safest storage racks started 15 months before the first rack installations were begun. I had already performed the first real seismic analysis on the rerack at Fort Calhoun, including third party review. Early in the '70's I was assigned the job of writing the QA program for Beaver Valley Plants. This document was the best QA program written for the nuclear power industry. When Westinghouse saw it they had fits. It was declawed and rewritten to please the reactor supplier. Unfortunately, this document was leaked to the AEC and became the basis for the government's requirement document for nuclear power plants. The words that scared Westinghouse the most were "traceability of all calculations of seismic analyses, thermal hydraulic analyses and nuclear analyses." (My Bettis training taught me to take nothing for granted.)

Wachter Associates initiated a seismic test program with Dr. Scavuzzo, head of the Mechanical Engineering Department at the University of Akron and a seismic consultant to the NRC. This program lasted five years and provided real time data and information for <u>engineering</u> analysis of spent fuel racks subjected to seismic events. (I was also an expert in seismic analysis based on the work at Bettis on the light water breeder reactor originally planned for California.) A series of tests meeting the criteria were performed. The results of these tests demonstrated how ridiculous the NRC criteria were. The NRC requirements document ignored "oil canning" of flimsy fuel cell walls, expected all earthquakes to occur in the east-west or north-south direction (ignoring the diagonal direction which is the weakest direction for the high density checkerboard racks with no top end stiffeners; they tend to deflect into rhombic shapes). The final and the most unrealistic requirement was to determine how far the racks moved after eight earthquakes. One of our first tests was rack motion during an earthquake. If the floor of the spent fuel pool was perfectly flat, the motion of the rack during an earthquake was in the range of 0.1 inch. If the pool sloped 0.002 inches per foot the rack slid down the slope until it impacted an obstruction. Too much of the NRC requirements were based on analyses based on virtual reality (professors) and not on the real world. We installed all our racks with zero gap which is where they will end up after the first earthquake. The bad thing about allowing gaps between racks was the effect of cumulative gaps as five or six rows of racks moved down a slope and hit. Most pools sloped toward the middle. Using the worst case scenario dictated by the NRC, the first down slope racks could be damaged by the impact loads imposed by rows of racks piling up on them.

In support of the tight packed racks, tests showed that these racks were dynamically coupled and acted as one rack driven by the dominant racks (heaviest load) in the array. This further supported and validated our analyses although we had long battles with the NRC to get our approach approved even though everyone else was <u>wrong</u>. (Not the first time I've had to battle for reality against virtual reality.)

The creation of the Wachter racks was based on years of development and analyses supported by tests performed at the Advanced Design Engineering Laboratory at the University of Akron. It took over a year to develop the process for making long boxes accurate to within ±0.015 inch in all dimensions. The sheared sheets used to make the boxes had to be ten times more accurate than the standard sheared sheets. I visited a sheet mill and observed the shearing process. Talking to workers revealed the ±0.030 inch tolerance on sheet width included the wear on the shearers. If they sharpened and aligned the shears frequently they could hold the accuracy of the sheet width to ±0.005 inch. Once we placed the requirement on our orders we could get the accuracy we needed to make accurate boxes. Fred Klingener established the minimum sheet thickness to avoid non-linear "oil canning" of box walls. For the PWR box (9" x 9") the thickness was 0.090"; for the BWR box (6" x 6") it was 0.075" thick. Fuel handlers have complained about the modern racks that have violated these criteria. They have experienced oil canning of fuel cell walls during fuel handling.

The problem of joining the boxes to form racks was solved by Jim Weissburg and Sid Kaufman. They came up with the tweezer welder. The Wachter racks proved to be eight to ten times as strong in the critical high stress areas as the racks furnished by other suppliers. The greatest weakness in the <u>other</u> racks was weakness in the diagonal direction and structural non-linearity. In 1978 Wachter Associates took 65% of all orders for spent fuel racks.

The last important element we developed for the <u>high</u> density racks using my invention of flux trap racks and poison wall racks, was the neutron poison panels forming the flux traps. The obvious choice was cadmium. (I used this material in the shipping box for transporting the S5W fuel assembly to B&W Lynchburg.) The only source in the United States for cadmium was a company in Brooklyn. I called them and asked about supplying this material. The man I talked to asked how much I would need. I said about 20,000 pounds. There was a long silence, then he said "I guess Elmer will have to work Saturdays." That was the end of my cadmium search. The French copied my early efforts and located a major source of sheet cadmium in their country. This is their preferred neutron poison after disastrous experiences with Boral swelling. (Boral finally became the U.S. preferred poison.) My early contacts with the people supplying Boral was very unsatisfactory. They were very unprofessional and refused to perform any irradiation tests to verify their product's ability to perform. This led to my formulation of Boroflex using boron carbide as poison mixed in powder form with silicon rubber. The silicon rubber was the most environmentally stable material available. But unfortunately, it required skill gained from experience with elastomers to handle it. I located (to my sorrow later), a company called Bisco to make this product. Dr. Franz of our company performed a total dose analysis for 20 and 30 years exposure to spent fuel. This was the first and best analysis performed on spent fuel exposure. A program of irradiation tests was initiated at the University of Michigan to verify the ability of Boroflex to take this amount of irradiation. The product was in production under the vigilant control of Wachter Associates. There were two major problems with this material, first the boron carbide had to be thoroughly dried (two to three furnace treatments) and second, the silicon rubber matrix had to be completely cured. After getting four fixed price orders based on the contract price Bisco gave me, they raised the price 40% and like a fool I swallowed the loss. This made Jim Sherwood, the engineer in charge of making Boroflex, mad so that he quit and offered to set up a company to make Boroflex for me. I unfortunately failed to take his offer. Not only was Bisco greedy, but they took short cuts in the manufacturing process that led to failures in later application of this product.

The first job for Wisconsin Electric, Point Beach reactors, went well and the Boroflex was up to our quality standards for the modulus of elasticity and strength. (The later material failed to meet these standards due to failure to completely cure the silicon rubber, and bad handling procedures by our competitors.)

At the time we were awarded the rack contract at Point Beach 1 and 2 (Wisconsin Electric), I had a meeting with Glenn Reed, VP of Nuclear Projects. (He was considered the Admiral Rickover of commercial nuclear power.) After the meeting Glenn and I went to the men's room. While standing at the urinals he said, "I wish I could buy two more nuclear plants from Westinghouse for $150 million each." Later he said, "I don't want any of that Boral used in these racks. When we installed the racks at Yankee Rowe, the stuff fizzed like Alka Seltzer." We had no other choice but Boroflex. It proved to be excellent material at Point Beach. We examined the 20 year test samples of Boroflex and they showed no change. (No breaks in the Boroflex sheets.)

William Wachter and John Nevshemal

All Wachter flux trap racks built at the same period as Point Beach had removable flux assemblies to make replacement of the nuclear poison material <u>easy</u>. We didn't trust any of the available nuclear poison materials in use at this time. No one else in the industry was smart enough to have removable poison designs.

CHAPTER 12
Design Safety in Nuclear and Non-Nuclear Systems

During the time I worked in nuclear power, people called on me to review a lot of non-nuclear systems. The general mechanical design errors causing costly lack of performance in the nuclear industry are common in all industry and government operations and hardware procurement. The same design mistakes were also discovered throughout the non-nuclear industry. No area of design in the world was free of stupid mistakes, showing an overall lack of basic common sense. Product reliability is dependent on testing to prevent otherwise disastrous failures from occurring.

As the main Mechanical Design and Reliability Consultant to Budd Manufacturing, Wachter Associates was in the role of reviewer and solver of all problems in the manufacture of the Metroliner engines and cars. At the opening meeting we visited the shops which made the large panels, girders and heavy components of both engine and car parts. We next visited the major assembly shop where complete electric engines were final-assembled. The most exciting part of the assembly process was the large network of electric wires and their connections. We recommended standard electronic wiring, color coded so that each wire was identified and traceable. Unfortunately, we found the condition of the assembly area to be quite poor. Workers were making electrical assemblies in a poorly lit area. We asked one worker about the accuracy of this work. He said the error rate was about thirty percent. After we had the lighting problem solved, this error rate was dramatically reduced to one half percent. Finally, it was discovered that due to bad design it would be necessary to use four Metroliner trains as spare parts just to keep the other four trains operating.

The next job we had was to ride the trains during initial operation. The rail system was just as bad as the trains. Westinghouse, which supplied the main electric power systems, made a major error caused by computer-controlled stupidity, a common problem throughout the engineering world paralyzed by computer control. The train's electric power system was not constant which caused the electric power sources to fail. Once we explained that they were not dealing with a constant power electric system, our redesign solved the problem. On the train, failures of GE equipment were so prevalent that GE engineers carried replacement parts on the trains in order to keep them running. I solved the problems by designing mountings for the electric units which eliminated the violent vibrations of the train. When all the work was done, the trains were able to do the job,

The Twentieth Century, The Age of Mediocrity

but as predicted, half of the trains were cannibalized (for parts) to keep the other four trains operating.

Shortly thereafter the Three Mile Island incident occurred. This had a disastrous effect on the United States' nuclear energy future. The prelude to a disaster states. In training the sailors of the nuclear navy, we gave them a solid education in the operation of a nuclear plant. An important part of the training was the use of all tools associated with the operation and maintenance of the mechanical-electrical systems. The cause of this accident can be placed on the operators and their poor training. A valve malfunction, which was automatically handled by the system, was interfered with by a poorly trained operator who disrupted the automatic safety systems. This led to disaster.

To compound this tragedy of errors was the fact that the designers of the two identical power plants, TMI-1 and TMI-2, saw fit to locate the critical visual instruments on the TMI-2 plant on the back side of the instrument panels where the operator could not see them. The TMI-1 plant did not misplace these instruments. The engineers who designed the plan were the worst contributors to this accident. They had two identical power plants, yet the control room of one plant was entirely different from the other, which added to the confusion of figuring out how the accident occurred. The greatest irony of all was that if all the operators had been expelled from the operating room at the time of the accident, the plant would have automatically shut down with <u>no</u> reactor operator action.

I have always known the "age of stupidity" was not restricted to the U.S. The Chernobyl incident was a demonstration of the combined stupidity of Russia, England and the U.S., resulting in the world's worst peacetime nuclear accident, killing thousands of people and contaminating Eastern Europe.

This nightmare occurred while I was at a nuclear meeting in Washington with people from Great Britain. It was quickly determined that this was an overheating without nuclear reaction, exactly like the fire at Windscale a few years before this accident. The Brit at the meeting had been at Windscale and was involved in putting out the fire which was non-nuclear. A dozen fire hoses put out this graphite blaze; no reactivity was released. U.S. delegates wanted me to go to Russia and put out the fire. I turned to the Brit and said that Britain should send experts from Windscale to help the Russians.

After the meeting, the Brits did nothing for 24 hours. I raised heck with the British Embassy, but got nowhere. I called the Russian nuclear engineer at their embassy and offered my services, adding my concern that this incident, if mishandled, could result in major fires that would create clouds of radioactive gases. The Russian said, "Ve are goink our own vey." I told him they would never get there and they never did. The fire burned for days and contaminated Europe. It also irradiated and killed operators of the Chernobyl plant. The combined idiocy of <u>three</u> countries is to blame for this mess.

The final fiasco was the Twin Towers disaster that was erroneously blamed on terrorists and gave them the biggest propaganda boost and did more to promote terrorism than any incident in recent history. The Twin Towers, designed by New York City architects, were two flimsy buildings which were held up by 5/8" boards that I wouldn't use on patio chairs. The failure of bolts holding up beams on one floor caused its collapse to the next floor initiating continuous collapse of floors all the way down, destroying the buildings and the people in them. The planes or any seismic or storm blast would have had the same result. I sent a letter to Tom Ridge to provide solutions to the building collapse program without regard to how flimsy they were. These were two inventions – the first strengthened the buildings so that they would not collapse in a heap, similar to that which occurred as a result of the Twin Towers disaster, and the second was a weapon that destroyed the planes before they could damage the buildings.

I also attended the first meeting, in Washington, D.C., to take action against terrorist attacks to which inventors were specifically invited. An engineering advisor to the President sent me a personal invitation. At the meeting the inventors were ignored. In fact, the Bureau of Standards, which hosted the meeting, informed me that the Patent Group had been disbanded since no one had replaced my friend Dr. Rabinean, (who was head of the Patent Group), since his death. The most ridiculous thing I ever heard of! Part way through the meeting I left, saying, "Call me back when you do something besides building tables for people to hide under." Another case of gutless mediocrity taking over a part of our government. I later sent another letter to Ridge about the safeguarding of buildings against airplane attacks. There was no reply.

THE AUTHORS' SUMMARIES OF QUALIFICATIONS

William Wachter

Mr. Wachter is a principal inventor of the SSW nuclear reactor used in shipboard applications and of safe storage systems for spent nuclear fuel from nuclear electric generating stations. The development of shipboard applications for nuclear power with his designs has led to the only nuclear power plants that have been produced in quantity in the USA. Later as a consultant, he saw a need to provide safe and efficient storage of highly irradiated spent nuclear fuel from commercial nuclear power plants.

His reactor inventions have had a positive material impact on the defense systems of the United States during the "cold war" and his inventions in spent fuel storage systems have helped to achieve safe long-term storage. These inventions have also had a significant impact on the Pittsburgh area because of the engineering efforts and the manufacturing of products generated by these inventions. Today nuclear power is accepted as standard for major naval vessels and over 27,000 spent fuel cells are safely stored in spent fuel racks of Mr. Wachter's design. Ionics, Inc. currently offers fuel racks of local manufacture with a recently patented Wachter design.

Mr. Wachter earned a Bachelor of Science Degree in Mechanical Engineering from the University of Pittsburgh in 1946 and a Master of Science Degree from Case Institute of Technology in 1955. Mr. Wachter is a member of the American Society of Mechanical Engineers and of the American Nuclear Society. He resides in Wexford, Pennsylvania, where he continues to conduct his consulting business, Wachter Associates. So far, he has thirty U.S. patents in addition to numerous patent applications held under Secrecy.

John Nevshemal

Mr. Nevshemal has extensive as well as unique experience in the analysis, operation, design, construction, modification/maintenance and D&D of nuclear facilities. The type of nuclear facilities include: commercial power plants; radwaste storage, processing and disposal; spent fuel storage; and research irradiation. This experience includes the supervision and administration of technical staff and engineers. Several activities included the direct development of safety analysis documentation for the modification and operation of radioactive waste storage and processing facilities. Also included is the determination of source terms, radiation shielding, personnel dose, hazards, accident scenarios, risk and consequences.

He is recognized both nationally and internationally for technical contributions to ANSI nuclear design standards, IAEA safety guides and technical bulletins on radwaste, high-level waste and spent nuclear fuel management. Also, he was the Chairman of ANSI/ANS-N48 consensus body for Restoration and Remediation of Radioactively Contaminated Sites and Radwaste

Management standards, Vice Chairman of ANSI/ANS-NUPPSCO consensus body for nuclear facility design standards and was the Chairman of ANSI/ANS Management Subcommittee ANS-55 for development and maintenance of nuclear fuel (new/spent) and radwaste handling, storage and processing standards. He has participated in the 10 CFR 50.59 Unreviewed Safety Question evaluation process at operating nuclear facilities. In his role as a Chief Nuclear Engineer, he was responsible for technical adequacy of nuclear projects as well as providing PEER review of system designs, shielding requirements and safety documentation.